CORTINA 1600E & GT 1967-1970

Compiled by
R.M. Clarke

ISBN 0 946 489 114

Distributed by
Brooklands Book Distribution Ltd.
'Holmerise', Seven Hills Road,
Cobham, Surrey, England

BROOKLANDS ROAD & TRACK SERIES

Road & Track on Austin Healey 1953-1970
Road & Track on Corvette 1953-1967
Road & Track on Corvette 1968-1982
Road & Track on Ferrari 1968-1974
Road & Track on Ferrari 1975-1981
Road & Track on Fiat Sports Car 1968-1981
Road & Track on Jaguar 1974-1982
Road & Track on Lamborghini 1964-1982
Road & Track on Lotus 1972-1981
Road & Track on Maserati 1952-1974
Road & Track on Maserati 1975-1983
Road & Track on Mercedes Sports & GT Cars 1970-1980
Road & Track on Porsche 1972-1975
Road & Track on Porsche 1975-1978
Road & Track on Porsche 1979-1982

BROOKLANDS MUSCLE CARS SERIES

American Motors Muscle Cars 1966-1970
Camaro Muscle Cars 1966-1972
Capri Muscle Cars 1969-1983
Chevrolet Muscle Cars 1966-1971
Dodge Muscle Cars 1967-1970
Mini Muscle Cars 1961-1979
Plymouth Muscle Cars 1966-1971
Muscle Cars Compared 1966-1971
Muscle Cars Compared Book 2 1965-1971

BROOKLANDS MILITARY VEHICLES SERIES

Allied Military Vehicles Collection No. 1
Jeep Collection No. 1

BROOKLANDS BOOKS SERIES

AC Cobra 1962-1969
Alfa Romeo Spider 1966-1981
Armstrong Siddeley Cars 1945-1960
Austin Seven 1922-1982
Austin 10 1932-1939
Austin A30 & A35 1951-1962
Austin Healey 100 1952-1959
Austin Healey 3000 1959-1967
Austin Healey 100 & 3000 Collection No. 1
Austin Healey Sprite 1958-1971
Avanti 1962-1983
BMW Six Cylinder Coupés 1969-1975
BMW 1600 Collection No. 1
BMW 2002 Collection No. 1
Buick Cars 1929-1939
Cadillac in the Sixties No. 1
Camaro 1966-1970
Chrysler Cars 1930-1939
Citroen Traction Avant 1934-1957
Citroen 2CV 1949-1982
Cobra & Replicas 1962-1983
Cortina 1600E & GT 1967-1970
Corvair 1959-1968
Corvette Cars 1955-1964
Daimler Dart & V-8 250 1959-1969
Datsun 240z & 260z 1970-1977
De Tomaso Collection No. 1
Dodge Cars 1924-1938
Ferrari Cars 1946-1956
Ferrari Cars 1957-1962
Ferrari Cars 1962-1966
Ferrari Cars 1966-1969
Ferrari Cars 1969-1973
Ferrari Cars 1973-1977
Ferrari Cars 1977-1981
Ferrari Collection No. 1
Fiat X1/9 1972-1980
Ford GT40 1964-1978
Ford Mustang 1964-1967
Ford Mustang 1967-1973
Ford RS Escort 1968-1980
Hudson & Railton Cars 1936-1940
Jaguar (& S.S) Cars 1931-1937
Jaguar (& S.S) Cars 1937-1947
Jaguar Cars 1948-1951
Jaguar Cars 1951-1953
Jaguar Cars 1955-1957
Jaguar Cars 1957-1961
Jaguar Cars 1961-1964
Jaguar Cars 1964-1968
Jaguar E-Type 1961-1966
Jaguar E-Type 1966-1971
Jaguar E-Type 1971-1975
Jaguar XKE Collection No. 1
Jaguar XJ6 1968-1972
Jaguar XJ6 1973-1980
Jaguar XJ12 1972-1980
Jaguar XJS 1975-1980
Jensen Cars 1946-1967
Jensen Cars 1967-1979
Jensen Interceptor 1966-1976
Jensen-Healey 1972-1976
Lamborghini Cars 1964-1970
Lamborghini Cars 1970-1975
Lamborghini Countach Collection No. 1
Land Rover 1948-1973
Land Rover 1958-1983
Lotus Cortina 1963-1970
Lotus Elan 1962-1973
Lotus Elan Collection No. 1
Lotus Elan Collection No. 2
Lotus Elite 1957-1964
Lotus Elite & Eclat 1975-1981
Lotus Esprit 1974-1981
Lotus Europa 1966-1975
Lotus Europa Collection No. 1
Lotus Seven 1957-1980
Lotus Seven Collection No. 1
Maserati 1965-1970
Maserati 1970-1975
Mazda RX-7 Collection No. 1
Mercedes Benz Cars 1949-1954
Mercedes Benz Cars 1954-1957
Mercedes Benz Cars 1957-1961
Mercedes Benz Competition Cars 1950-1957
MG Cars in the Thirties
MG Cars 1929-1934
MG Cars 1935-1940
MG Cars 1948-1951
MG TC 1945-1949
MG TD 1949-1953
MG TF 1953-1955
MG Cars 1952-1954
MG Cars 1955-1957
MG Cars 1957-1959
MG Cars 1959-1962
MG Midget 1961-1979
MG MGA 1955-1962
MGA Collection No. 1
MG MGB 1962-1970
MG MGB 1970-1980
MG MGB GT 1965-1980
Mini Cooper 1961-1971
Morgan Cars 1936-1960
Morgan Cars 1960-1970
Morgan Cars 1969-1979
Morris Minor 1949-1970
Morris Minor Collection No. 1
Nash Metropolitan 1954-1961
Opel GT 1968-1973
Packard Cars 1920-1942
Pantera 1970-1973
Pantera & Mangusta 1969-1974
Pontiac GTO 1964-1970
Pontiac Firebird 1967-1973
Porsche Cars 1952-1956
Porsche Cars 1960-1964
Porsche Cars 1964-1968
Porsche Cars 1968-1972
Porsche Cars in the Sixties
Porsche Cars 1972-1975
Porsche 356 1952-1965
Porsche 911 Collection No. 1
Porsche 911 Collection No. 2
Porsche 914 1969-1975
Porsche 924 1975-1981
Porsche 928 Collection No. 1
Porsche Turbo Collection No. 1
Reliant Scimitar 1964-1982
Riley Cars 1945-1950
Riley Cars 1950-1955
Rolls Royce Cars 1930-1935
Rolls Royce Cars 1940-1950
Rolls Royce Silver Cloud 1955-1965
Rolls Royce Silver Shadow 1965-1980
Range Rover 1970-1981
Rover 3 & 3.5 Litre 1958-1973
Rover P4 1949-1959
Rover P4 1955-1964
Rover 2000 + 2200 1963-1977
Saab Sonett Collection No. 1
Saab Turbo 1976-1983
Singer Sports Cars 1933-1954
Studebaker Cars 1923-1939
Sunbeam Alpine & Tiger 1959-1967
Triumph Spitfire 1962-1980
Triumph Spitfire Collection No. 1
Triumph Stag 1970-1980
Triumph TR2 & TR3 1952-1960
Triumph TR4 & TR5 1961-1969
Triumph TR6 1969-1976
Triumph TR7 & TR8 1975-1981
Triumph GT6 1966-1974
Triumph Vitesse & Herald 1959-1971
TVR 1960-1980
Volkswagen Cars 1936-1956
VW Beetle 1956-1977
VW Beetle Collection No. 1
VW Karmann Ghia Collection No. 1
VW Scirocco 1974-1981
Volvo 1800 1960-1973
Volvo 120 Series 1956-1970

CONTENTS

ACKNOWLEDGEMENTS

Motor companies in general and Ford in particular have had their marketing failures and successes on both sides of the Atlantic. In the US they got it wrong with the Edsel and then scooped the pool a few years later with the Mustang. In Britain there was the painful episode with the Consul Classic, followed shortly afterwards by the Cortina which has probably been the most profitable model ever produced by Ford in Europe.

The Cortina was launched in 1962 and was subsequently developed with great marketing skill. A year later, the Lotus Cortina attracted much attention because of its brisk performance and went on to gain many successes on the track. A bold decision in 1966 was to completely rebody the car which proceeded to distance it from its main competitor of the period – the Austin/Morris/MG 1100.

Ford made two significant marketing moves in 1967. They gave the top models better acceleration and a higher maximum speed by increasing the engine capacity to 1600cc, then gilded the lily by introducing an Executive Cortina. This bestowed on the marque the prestige it had lacked up to that time. The 1600E achieved its aims and has gone on to become a 'recent classic' with a large following which now supports an active owners club.

Brooklands Books cater to the enthusiasts who restore and cherish collectible cars such as the 1600E and GT, by making available the stories which were written about their vehicles when they were in production. This can only be done because the publishers of the world's leading motor journals generously allow us to include their copyright articles in this reference series.

We are sure that Cortina owners will wish to join with us in thanking the management of Autocar, Car, Car & Driver, Car Life, Motor, Motor Racing, Motor Sport, Old Motor, Practical Classics, Road & Track, Sports Car World, Thoroughbred & Classic Cars, and Wheels for their ongoing help, support and consideration.

R.M. Clarke

FORD CORTINA 1600E 1,599 c.c.

AT-A-GLANCE: Ford Cortina GT with luxury specification. Eager performance from 1600 cross-flow engine. Fuel consumption varies with driving. Brakes heavy for rapid stops; little fade. Steering, handling and gearbox first class. Roadholding good. Ride too hard. Unexpectedly noisy for its price.

MANUFACTURER:

Ford Motor Co. Ltd., Warley, Brentwood, Essex.

PRICES:

Basic	£799	0s	0d
Purchase Tax	£183	2s	1d
Total (in G.B.)	£982	2s	1d

PERFORMANCE SUMMARY

Mean maximum speed	98 mph
Standing start ¼-mile	18.8 sec
0-60 mph	13.1 sec
30-70 mph (through gears)	13.7 sec
Typical fuel consumption	27 mpg
Miles per tankful	270

ALL the multitude of accessories offered (and bought) for the embellishment of ordinary cars is enough to show that many buyers want, in the American phrase, to "personalize" their transport. In an age of quantity production and increasing conformity, it is understandable that more people should wish to be slightly different. Ford first acknowledged that need with the Executive Zodiac (now called simply Executive) and then with the Corsair 2000E; on the eve of this year's Motor Show they introduced the Cortina 1600E, which is basically a four-door Cortina GT with a number of "luxury" extras as standard. Ford's "executive" (for which the "E" stands) is reckoned as a successful young married businessman seeking a four-seater sporting saloon, with more comfort and status than is normal for this class of car.

Basic equipment includes therefore reclining seats, polished wood facia and window sills, an imitation leather-rimmed steering wheel, auxiliary lamps, automatic reversing lamps, a clock, a radio, extra storage space and even partial trimming inside the boot. The body has four doors and sits on the lower, Cortina-Lotus type suspension with wide (5.5 instead of 4.5in.) rim "sculptured" pressed-steel wheels, which having no nave plates are partially chromium plated. Tyres are now exclusively India Autoband radial-ply, 165-13 tubeless.

This is the first of the new cross-flow Cortinas we have tested, sharing with the GT the most powerful, 1,599 c.c. pushrod engine of the range. Its downdraught Weber twin-choke carburettor, "sporting" valve timing and tubular four-branch manifold combine to give it 88 bhp (net) at 5,400 rpm instead of the 78 at 5,200 for the previous 1,500 c.c. GT. Maximum torque has not increased so markedly—96 lb.ft. instead of 91 at 3,600 rpm. The previous Cortina GT tested by AUTOCAR was a 1500 (8 January 1965), which the 1600E beats in all but top gear acceleration from low speeds; it even out-performs the Corsair 2000E from standing starts.

Comparative figures for the 1600E, 1500 GT and 2000E respectively are as follows: 0-80 mph in 26.6, 31.1 and 27.7 sec; standing ¼-mile in 18.8, 18.7 and 18.8 sec; 30-50 mph in top gear 10.0, 9.6 and 9.4 sec; 60-80 mph in top gear 13.7, 18.1 and 14.5 sec. Larger tyres and a higher bottom gear than on the 1500 GT made it difficult to achieve really rapid starts on dry concrete, because they effectively eliminate all wheel spin.

Re-starting on the 1-in-3 test hill needed a lot of clutch slip and high rpm, but on the road the gear ratios are all that the keen driver could

Lowered suspension and fatter tyres give the basic Cortina body a thrusting look

FORD CORTINA 1600E . . .

ask for. The engine is very willing to spin up to the 6,000 red mark on the rev-counter and beyond, although it feels comparatively "flat" below about 3,500 rpm. The close and even spacing of the ratios (35, 51 and 73 mph maxima) make it easy to keep the engine "on full song" during hard driving. One seems to be invited to drive enterprisingly as soon as the unit is warm, which is soon after each very easy start.

Unfortunately anything above 4,700 rpm is unduly noisy inside the car. Back seat passengers notice a loud exhaust-induced boom, and the noise level is higher than is general for this class of car. As far as 70 mph-limited Britain is concerned, this is perhaps not so serious except in the indirect gears but high speed cruising on the Continent could be uncomfortable. An overdrive, which is not available on any Ford but the Zephyr V6 models, would be advantageous.

Cross-flow engine flexible

Although the top gear performance figures compared with the previous GT's do not suggest so, the cross-flow engine is more flexible. Pulling from 20 mph causes a momentary rocking of the engine, after which it runs smoothly. Fuel consumption varies between 21 mpg in town to nearly 30 on the open road, depending on how much one uses full throttle and the second choke of the carburettor. Steady-speed consumption is roughly 10 per cent heavier than for the 1500 GT; oil consumption for a Cortina was high at 650 miles per pint.

The gearchange is superb, having a narrow slot across its gate and crisp fore-and-aft movements with unbeatable synchromesh, yet pleasingly light action and no stickiness, apart from a slight initial "click". The clutch is similarly light and, as is usual with the modern diaphragm spring type, it is progressive and always smooth in taking up. Maximum braking, though extremely effective at 1.05g on dry asphalt, needs 120lb on the pedal, which is a lot these days. At 75 and 100lb we recorded good deceleration rates with the back wheels locked. The new handbrake worked by a T-handle under the facia proved extremely powerful and held easily on 1-in-3 with 0.37g retardation when used on its own from 30 mph. During our fade test, the pedal pressure for the final 0.5g stop from 70 mph had only increased by 30 per cent to just over 50lb.

The 1600E (like the Lotus on which it is based) has remarkably good steering, with little noticeable lost motion and a great deal of "feel" of the road. The 35ft 3in. mean turning circle with 4.25 turns lock-to-lock is not excessive and effort at the nice solid-rubber-cored steering wheel rim is only heavy when parking. From the high driving position overlooking a low nose, one seems to be very much in charge and able to place the car tidily and neatly on winding roads. Experience with previous Cortina GTs has taught us that the recommended radial-ply tyre pressures of 24 psi front —28 psi rear are better if changed to 28 psi all round; this cuts down the car's understeer without seriously affecting ride.

Normally this Cortina holds the road well, understeering slightly with little roll. Cornered hard in low gear it will eventually lift enough weight off the inside rear wheel to let it spin. If on a track it is thrown at a corner hard and fast some way before the apex, its tail will come out and stay out controllably under power; the steering is ideally geared for any correction necessary.

In complete contrast, the ride itself is not nearly so good. There is a pronounced vertical jogging on any sort of irregularity and all our testers felt it was too firm for a car in this class. Road noise is noticed both as bump-thump and tyre-roar on coarse surfaces, and there is considerable wind noise from the front door seals at speed. On our test car, a piercing hiss from the left-hand rear door seal made matters worse.

Front seats are somewhat hard, so that one seems to be sitting on, rather than in, them; there is not enough sideways location for a car which corners so well. At last a proper sliding adjustment is provided (instead of the previous stiff swinging linkage) but tall drivers cannot get far enough away from the pedals. Knee-room behind is cramped except when the front seats are right forward. The rear seat does not give enough support to the upper parts of the spine and there is no central armrest.

Engine accessibility is excellent. Note the efficient-looking four-branch tubular exhaust manifold

Autocar road test Number 216 Make: Ford Type: Cortina 1600E (1,599 c.c.)

TEST CONDITIONS: Weather: Fine, dry. Wind: 5-10 mph. Temperature: 10 deg. C. (50 deg. F.). Barometer: 29.55 in. Hg. Humidity: 60 per cent. Surfaces: Dry, concrete and asphalt.

WEIGHT: Kerb weight 18.4 cwt (2,064lb—937kg) (with oil, water and half-full fuel tank). Distribution, per cent: F, 54.3; R, 45.7. Laden as tested: 22.4 cwt (2,507lb—1,137kg).

Figures taken at 4,900 miles by our own staff at the Motor Industry Research Association proving ground at Nuneaton.

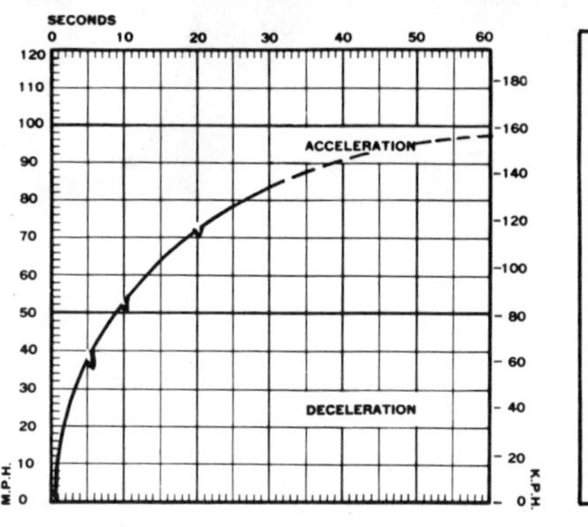

TIME IN SECONDS	0	4.1	6.1	9.1	13.1	17.8	26.6	38.8	
TRUE SPEED MPH		30	40	50	60	70	80	90	100
INDICATED SPEED		28	40	50	58	69	79	92	104

Mileage recorder 1.4 per cent over-reading
Test distance 1,963 miles.

MAXIMUM SPEEDS

Gear	Mph	kph	rpm
Top (mean)	98	158	5,750
(best)	100	161	5,850
3rd	73	117	6,000
2nd	51	82	6,000
1st	35	56	6,000

Standing ¼-mile 18.8 sec 72 mph
Standing Kilometre 35.5 sec 88 mph

Speed range, gear ratios and time in seconds

mph	Top (3.90)	3rd (5.45)	2nd (7.84)	1st (11.59)
10-30	—	8.2	5.2	3.2
20-40	11.5	6.4	4.8	—
30-50	10.0	6.3	4.5	—
40-60	10.4	7.0	—	—
50-70	12.1	8.1	—	—
60-80	13.7	—	—	—
70-90	19.4	—	—	—

FUEL CONSUMPTION

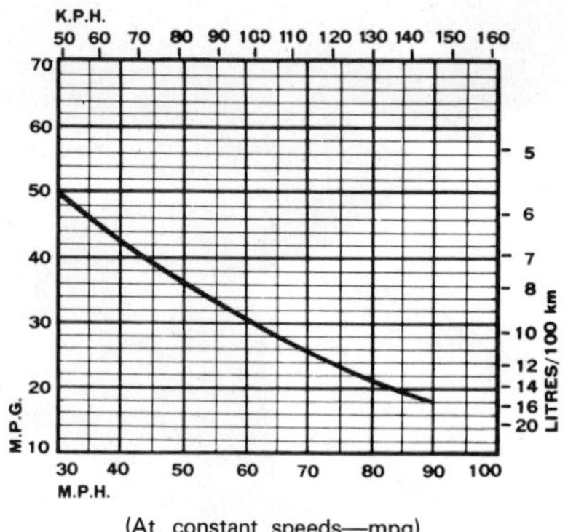

(At constant speeds—mpg)

30 mph	50.0
40	43.0
50	35.8
60	30.8
70	26.1
80	21.7
90	17.8

Typical mpg 27 (10.5 litres/100km)
Calculated (DIN) mpg . . 23.7 (11.9 litres/100km)
Overall mpg 25.1 (11.3 litres/100km)
Grade of fuel . . . . Premium, 4-star (min. 97RM)

OIL CONSUMPTION

Miles per pint (SAE 10W/30) 650

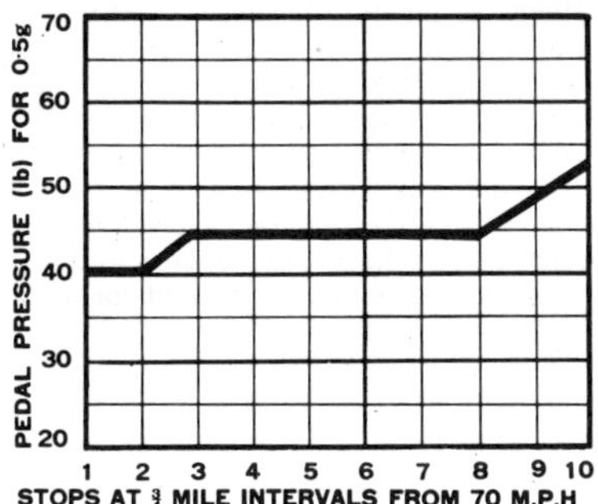

BRAKES (from 30 mph in neutral)

Load lb	g	Distance ft
25	0.23	131
50	0.50	60
75	0.80	38
100	0.95	32
125	1.05	28.7
Handbrake	0.37	81

Max. Gradient: 1 in 3

Clutch Pedal: 30lb and 5.25in.

TURNING CIRCLES

Between kerbs L, 33ft 1in.; R, 33ft 5in.
Between walls L, 35ft 1in.; R, 35ft 5in.
Steering wheel turns, lock to lock 4.25

OIL GAUGE
2 SPEED FAN
TEMPERATURE GAUGE
INTERIOR LIGHT
FUEL GAUGE
PANEL LIGHTS
INDICATOR TELL-TALE
AMMETER
REVOLUTION COUNTER
SPEEDOMETER
HEATER & DEMISTER
LAMPS
GLOVE BOX
INDICATOR TELL-TALE
SWIVELLING VENTILATOR
IGNITION & STARTER
SWIVELLING VENTILATOR
HORN DIPSWITCH & HEADLAMP SIGNALLER & INDICATORS
BONNET RELEASE
1 3
2 4 R
CHOKE
ASH TRAY
GLOVEBOX
CIGAR LIGHTER
CLOCK
SCREENWASH & WIPERS
HANDBRAKE
IGNITION LIGHT
MAINBEAM TELL-TALE

HOW THE CAR COMPARES:

Maximum Speed (mean) mph

70 80 90 100 110
Ford Cortina 1600E
BMW 1600
Fiat 125
Humber Sceptre
Vauxhall Victor 2000

0-60 mph (sec)

30 20 10
Ford Cortina 1600E
BMW 1600
Fiat 125
Humber Sceptre
Vauxhall Victor 2000

Standing Start ¼-mile (sec)

30 20 10
Ford Cortina 1600E
BMW 1600
Fiat 125
Humber Sceptre
Vauxhall Victor 2000

MPG Overall

20 30 40
Ford Cortina 1600E
BMW 1600
Fiat 125
Humber Sceptre
Vauxhall Victor 2000

PRICES

Ford Cortina 1600E	£982
BMW 1600 Coupé	£1,298
Fiat 125	£999
Humber Sceptre	£1,139
Vauxhall Victor 2000	£910

Left: The handsome polished aluminium-alloy steering wheel has a thick leather-covered rim; polished wood covers the entire dashboard. Reverse can now be selected without lifting the short gearlever. Right: Reclining seats are standard and with the front seats slid forward can be laid flat. The armrest in front is the lid of a useful compartment; there is no central armrest behind

FORD CORTINA 1600E . . .

All controls are within reach of a belted driver but longer switch arms would prevent the handsome thick wooden dashboard from having its varnish scratched. The speedometer proved to be accurate or very slightly low-reading up to 80 mph, while the mileometer over-read by 1.4 per cent; it seems a pity that there is no trip recorder on a car of this price. Windscreen wipers sweep a generous area with little glass uncleared, but they have only one speed. The interior mirror is flat and does not show enough of the rear window, while the two standard mirrors are also flat and too restricted in their field of view.

Safety protection in a crash has been looked after with a padded roll on top of the facia, the dished steering wheel, crushable vizors and parcel shelf and a degree of padding inside the roof lining.

On all latest Cortinas the Aeroflow heating and ventilation system has been slightly simplified by doing away with the temperature control for each swivelling "eyeball". There are now butterfly valves in each vent which control the flow of cold air only. Ford argue that warm air is rarely wanted from these. Certainly we found that even after one icy night with frost over all windows, it was only the back screen which had not started to defrost after only four minutes fast idle from cold. Indeed, the ice on the windscreen had melted and started to steam at the end of that time. The only snag seems to be that the butterflies sometimes do not always seal completely and one feels a slight cold draught on one's knees. Otherwise the system remains an object lesson to all other manufacturers on how to make winter motoring so much more comfortable.

There is plenty of space in the partially carpeted boot and the spare wheel under its cover is neatly tucked behind one wheel arch. Loading is not easy because of the high sill and there are mixed feelings about the lock having to be worked with the key each time the lid is opened. Inside the car, the child safety locks on the rear doors have now been improved and can be worked without the need for a coin or implement.

Below: 1600E owners will already know how much attention these racey-looking pressed steel wheels attract. It is essential to clean them regularly during snowy weather to avoid salt corrosion of the brightwork. Bottom: Partially carpeted boot space is generous and the spare wheel tucks neatly behind one wheel arch. Right: Heavy polished wood door cappings lend a luxury look to the interior. Rear legroom is limited for tall passengers

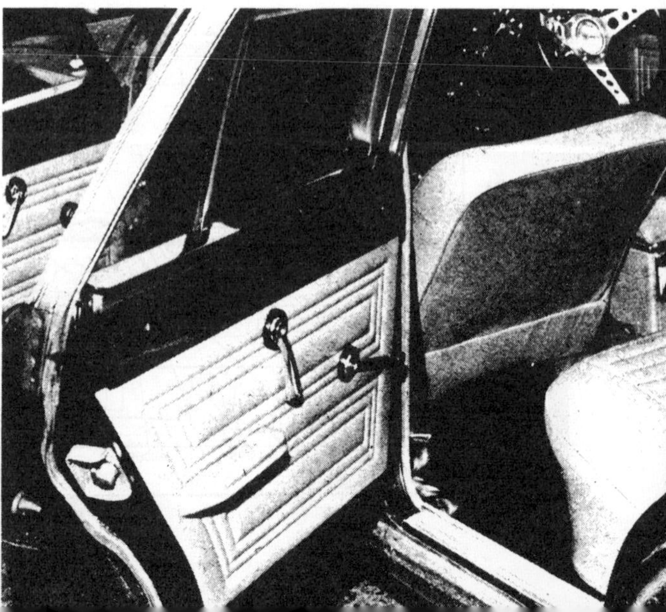

As on all Cortinas the 10-gallon fuel tank demands careful filling otherwise it spits back messily; also a larger tank would be appreciated on long trips. The headlamps are combined with two long-range auxiliaries which come on automatically with main beams, making in effect a four-headlamp system.

Generally speaking, accessibility to items requiring maintenance is very easy since the neat little engine is relatively lost in the wide spaces under the bonnet. The Cortina body has few traps for the fingers of anybody washing it.

Despite its earlier-conceived "executive" brothers, the 1600E is a very enterprising transformation of a popular medium saloon. It is undoubtedly great fun to drive but it lacks some of the refinement one expects from a car costing nearly £1,000. As far as equipment goes there are few items one might wish to add. Many will be glad of the excuse to buy this Ford without the loaded insurance and "racer" image of the stablemate from which it inherits so many of its finer points. □

SPECIFICATION: FORD CORTINA 1600E (FRONT ENGINE, REAR-WHEEL DRIVE)

ENGINE

Cylinders	4, in line
Cooling system	Water; pump, fan and thermostat
Bore	81.0mm (3.19in.)
Stroke	77.6mm (3.06in.)
Displacement	1,599 c.c. (97.6 cu. in.)
Valve gear	Overhead, pushrods and rockers
Compression ratio	9.0-to-1: Min. octane rating: 97RM
Carburettor	One Weber 32 DFM compound twin choke downdraught
Fuel pump	AC mechanical
Oil filter	Fram or Tecalemit full flow, renewable element
Max. power	88 bhp (net) at 5,400 rpm
Max. torque	96 lb.ft. (net) at 3,600 rpm

TRANSMISSION

Clutch	Borg and Beck single plate diaphragm spring 7.54in. dia.
Gearbox	Four-speed, all-synchromesh
Gear ratios	Top 1.0 Third 1.40 Second 2.01 First 2.97 Reverse 3.32
Final drive	Hypoid bevel, 3.90-to-1

CHASSIS and BODY

Construction	Integral, with steel body

SUSPENSION

Front	Independent, MacPherson struts, coil springs, anti-roll bar, telescopic dampers
Rear	Live axle, half-elliptic leaf springs, radius arms, telescopic dampers

STEERING

Type	Burman recirculating ball
Wheel dia.	15.25

BRAKES

Make and type	Girling disc front, drum rear
Servo	None
Dimensions	F. 9.63in. dia.; R. 9.00in. dia.; 1.75in. wide shoes
Swept area	F. 189.5 sq.in.; R. 98.9 sq.in. Total 288.4 sq.in. (257.5 sq.in./ton laden)

WHEELS

Type	Pressed steel disc, 5.5in. wide rim
Tyres—make	India
—type	Autoband radial-ply tubeless
—size	165-13 mm

EQUIPMENT

Battery	12-volt 38-Ah
Generator	Lucas C40 22-amp d.c.
Headlamps	Lucas sealed filament, twin Wipac spot lamps on main beam 216/90-watt (total)
Reversing lamp	Standard
Electric fuses	None
Screen wipers	Single speed, self-parking
Screen washer	Standard, manual plunger
Interior heater	Standard, air-blending type
Heated backlight	Not available
Safety belts	Standard
Interior trim	Pvc seats, pvc headlining
Floor covering	Carpet
Starting handle	No provision
Jack	Screw pillar
Jacking points	Two each side under sills
Windscreen	Zone toughened
Underbody protection	Phosphate treatment prior to painting

MAINTENANCE

Fuel tank	10 Imp. gallons (no reserve) (45.5 litres)
Cooling system	11.4 pints (including heater)
Engine sump	6.2 pints (3.5 litres) SAE 10W/30. Change oil every 6,000 miles. Change filter element every 6,000 miles
Gearbox	2.1 pints SAE 80EP. Change oil at first 3,000 miles only
Final drive	2 pints SAE 90EP. No oil change needed
Grease	None needed
Tyre pressures	F. 24; R. 28 p.s.i. (normal driving) F. 28; R. 28 p.s.i. (fast driving)

PERFORMANCE DATA

Top gear mph per 1,000 rpm	17.1
Mean piston speed at max power	2,750 ft./min.
Bhp per ton laden	78.6

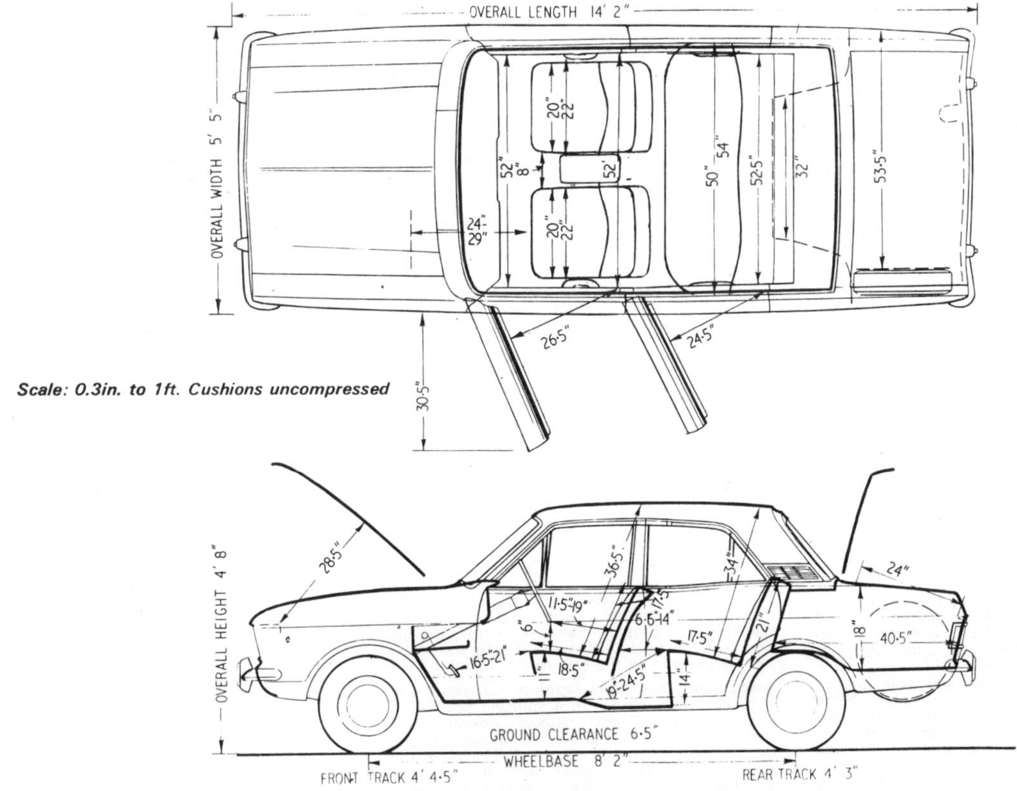

Scale: 0.3in. to 1ft. Cushions uncompressed

Ford Cortina 1600E

Description: The latest 'Crossflow' Cortina GT 'executive-ised' to produce a fast and luxurious vehicle for under £1,000. The idea behind the Executive Fords is apparently to build in so much luxury as standard that there isn't much room left for optional extras. In the case of the brand new Cortina, as opposed to the earlier Zodiac and Corsair models, the accent has changed slightly to produce a car with really sporty appearance, handling and performance.

Engine and transmission: Ford's new 1968 'Crossflow' engine is coupled to the usual excellent all-synchromesh four forward speeds and reverse gearbox, though an automatic transmission is an optional extra. The engine has over-square bore and stroke measurements of 80.978 mm x 77.62 mm, giving a capacity of 1,598 cc. With bowl-in-piston combustion chambers and the new cross-flow cylinder head, the GT engine has a compression of 9.0:1 and produces 93.0 bhp at 5,400 rpm. A twin-choke Weber carburettor is fitted, and a long four-branch exhaust manifold is standard. Drive is taken via a 7.54 inch diameter self-adjusting diaphragm clutch to that beautiful gearbox, and thence to the hypoid rear axle with its 3.9:1 final drive ratio.

Suspension and brakes: The Lotus-Cortina's lowered suspension is fitted to the 1600E, with an independent system at the front by Macpherson telescopic damper struts and co-axial coil springs and anti-roll bar, and a rigid axle at the rear suspended on semi-elliptic springs and located by twin trailing links. Very attractive chrome and black 'sculptured' pressed steel 5½J wide-rim wheels are used all round (as on the Rovers 2000 and 3.5 V8), and India Autoband radial ply tyres were fitted to our test car. Something very peculiar has happened with Ford's offering tyres as standard, for both India and Pirelli have advertised their products as the 1600E's standard wear, and Cinturatos were fitted to the car at Earls Court! Brakes are the usual disc and drum assortment, with 9.62 inch diameter discs at the front and 9 inch diameter drums at the rear. An umbrella handle-type hand brake is fitted beneath the facia to the left of the steering column.

Accommodation and fittings: It is the interior of this car which has altered most, with fully reclining front seats, a polished wood facia and waist rails on the four doors (real tree wood!) and a leather-covered, only slighly dished aluminium spoked steering wheel. A quite long leather-gaited gearchange lever is mounted on the transmission tunnel centre console, and this unit also incorporates a clock and, between the seats, a thickly padded arm rest which lifts to open a deep storage compartment. Extra sound insulation has been fitted to the car, and the black cut pile floor carpeting also helps to deaden external noises. Other additions to the standard GT specification are dual-tone horns, a cigar lighter on the facia, and a spare wheel cover for this item which stands on one side of the spacious plastic carpeted boot. Externally the 1600E is very attractive, and in addition to the cast wide-rim wheels and lowered suspension has a matt black grille mounting powerful twin driving lamps as well as the normal headlights, a thin cheat line along either side (on our metallic light blue test car this was in dark blue) and '1600E' boot lid and rear quarter panel badges. Automatic reversing lights mounted beneath the back bumper are also standard. Upholstery was in cream throughout, making for a very light and attractive (though perhaps not terribly practical) interior.

Findings: This car is a knock-out. There are things wrong with it, but in general its standard of comfort and what our American friends call 'roadability' make it a delightful road car.

Criticisms first: The wooden facia has small holes drilled in it to accept some of the flick switches which are mounted on the metal underneath, and because of this the thickness of the facia has tended to bury the switches, and in particular the main lighting control. This means groping for it in the dark, and is a feature which could easily be improved by mounting on, rather than under, the wooden trim. The brakes were also rather spongy, and did not give any great degree of confidence, which was surprising for this Cortina GT system is usually pretty good, and with those big boots should have been even better than normal. Our test car was pretty new, though, and it may have been a question of bedding-in on those front discs. One or two annoying rattles appeared during the test—spoiling the luxury air somewhat.

But apart from these points the 1600E was very difficult to fault, particularly when bearing in mind the cost of just £982 2s 1d. The deep and luxurious seats, the powerful heating and demisting Aeroflow system, the very low interior noise level, delightful gearbox and most predictable handling qualities add up to one of the most impressive cars we have tested this year. We had our doubts as to the efficiency of the India Autoband tyres at first—ignorance breeding contempt — but they proved to grip well, give a smooth ride and have quite progressive breakaway characteristics. Just like the Lotus-Cortina, the 1600E understeers slightly before the tail takes over in a gentle and easily contained slide and, though maybe our memory is playing tricks here, its steering seems slightly more positive than that of the Lotus. There was very little body roll on the lowered suspension, which was stiff enough to feel slight bumps but with the deep seat cushions gave only a sensation of being in stable contact with the road, as opposed to leaping from lump to lump like a mountain goat!

The 'Crossflow' engine was still slightly on the tight side, but still pushed the tachometer needle round the clock to the 5,500 rpm red line with commendable speed, and on the move its mid-range torque is invaluable for sorting out 70 mph motorway bunches, giving vivid acceleration. Away from rest the figures we recorded were not as good as Ford claim, but ours were taken with a heavyweight crew aboard and, as already mentioned, the engine was not perhaps as free as it should be. The gearbox was delightful, with rather a long lever movement but faultless synchromesh on all four combining with the light diaphragm spring clutch for very quick cog-swopping. Power on in tight corners could spin the rear wheels and kick the tail round, but otherwise it is well and truly tacked down.

0-30 mph	4.5s
0-40 mph	6.9s
0-50 mph	10.0s
0-60 mph	13.5s (12.5s claimed)
0-70 mph	16.7s

Top speed is around 96 mph on the flat, though an indicated 100 mph is possible downhill, with the tachometer needle right on the red line. Touring fuel consumption is quoted at 31.5 mpg, but over 300 quite hard miles in our hands returned only 24 mpg—but a lot of town driving was included here. Just over 29 mpg was given by a gentle run in the country.

Assessment: Not quite faultless, but nearly so. For the family man with a sporting yen this is a 'grand touring sports saloon', offering great value for money, a high degree of comfort and general equipment and much of the speed and all of the safety of a good two-seat sports car. The best thing we've tried this year, and we can't say much more in its favour than that . . .

FORD CORTINA 1600 GT

Continued from page **30**

we see interesting variations in safety belt design. Most American cars use two separate belts—one for lap and one for shoulder. The Japanese favor a detachable shoulder belt. The Cortina uses just one strap with a sliding buckle to equalize tension between the lap and shoulder sections. The disadvantage with this system is that it puts the wearer in an all-or-none situation rather than allowing him an option of choosing the lap belt only.

Have we managed to say yet that the Cortina is a British car? Whenever we try to describe it we come back to that statement. It has some endearing characteristics, we loved to flog it because it feels strong and responsive although it never gives the impression of being a precision machine. The engine and drive train are very noisy contributing to a high background noise level at cruising speeds, the ride is quite harsh, particularly when we discovered that ride comfort hasn't been traded for good handling. Light cars normally have poor resistance to crosswinds and the Cortina is light *and* substandard for its weight.

Most of all, though, we were impressed with the GT's performance. The car is quick for a sedan of its displacement and price, almost as good as a BMW 1600 with a $350 less imposing list price. We're left with the conclusion that the Cortina GT is a hard-nosed little flogger car, and to hell with everything else. That's no bad thing to be. ●

For going-places people —the Cortina 1600E.

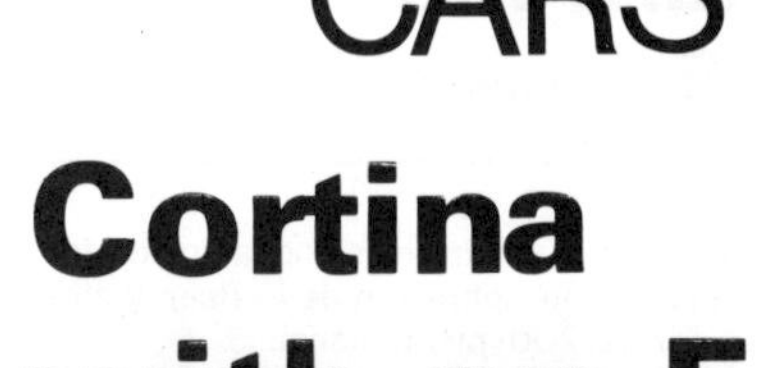

Cortina with an E

Executive version of 1600 and improvements to Corsairs, Zephyrs and Zodiac

The most luxurious Cortina interior yet. Note the wood facia, clock on centre console, and reclining seats.

The sporty leather-covered steering wheel, and remote-control gear-change are standard equipment.

AN Executive Cortina, de luxe versions of the Zephyr and Zephyr V6 range, refinements to the Corsair range, and detail but significant changes to the Zephyr and Zodiac models—that is the Ford line-up for the Show.

Encouraged by the success of the Executive version of the Corsair, Ford have amalgamated some of the most attractive features of the Cortina GT and the Cortina Lotus to make the Cortina 1600E. The new car is based very closely on the Cortina GT, but it is available only in the four-door version and is very much better equipped—in fact it features all the "extras" which we imagine a GT owner would want to fit.

Perhaps the most important modification is the lowered suspension, as on the Cortina-Lotus, and 5½J chromed "sculptured" wheels fitted with radial ply tyres. Twin driving lamps are mounted in the black-painted grille, automatic reversing lights are fitted, and other external distinguishing features include a contrasting "coach strip" along the waist-line of the car and special badges on the rear quarter panels.

The Luxury part of the Executive comes with fully reclining front bucket seats, a leather covered aluminium-spoked steering wheel and polished wood facia and door waist rails. Extra sound insulation and a black-cut pile carpet are standard, the gear lever has a leather gaiter, and the centre console unit features a thick, padded arm rest, storage space for small items, and a clock. Dual-tone horns, a cigar lighter and spare wheel cover make up the rest of the "mods".

Priced at £982 2s. 1d. including purchase tax, it fits into the Cortina range at almost £100 more than the four-door GT, and £90 cheaper than the Cortina-Lotus.

At the top end of the Ford line-up come de luxe versions of the Zephyr and Zephyr V6, each with radial ply tyres on 14-in. wheels—an inch larger than those on previous models. On all the Mk. IV models the camber angles on the independent rear suspension have been revised so that an appreciable degree of negative camber can now be induced; the steering ratio on models which don't have power steering has been raised from 20.6 : 1 to 23.5 : 1, and an improved servo unit is fitted to reduce braking effort.

The new Zephyrs are identified by a full-width extruded aluminium radiator grille and by a motif on the centre line of the bonnet. Individual front bucket seats are a standard fitting, but reclining seats are available at extra cost; between the seats is a console unit

Cortina with an E *continued*

Distinguishing features of the V6 de luxe are the new radiator grille, bonnet badge, and centre-bonnet motif.

providing a central armrest and a locker for small items. The upholstery is in deeply embossed Cirrus 200 pvc material.

Zodiac models now have the power steering which has been standard on the Executive, the grille badge previously seen only on the Executive, and the bonnet motif.

The de luxe Zephyr costs £961 4s. 2d., and the V6 de luxe £1,034 19s. 2d. (both prices include purchase tax).

Refinements on the Corsair range for next year include a walnut facia panel for the V4 de luxe. The car has additional safety features in a breakaway stem interior mirror, safety locks to prevent the front seats from tipping forward, a collapsible half-width parcels tray and a new and more powerful handbrake which is smoother in operation. The Corsair 2000E has the safety measures, and now also has a twin-bucket style rear seat; an electric clock is mounted on the centre console on manual transmission models. Both the 2000 de luxe saloon and the estate car now have servo assisted brakes and radial ply tyres in their basic specification.

On the Corsairs, Ford have replaced the four-position facia panel controls which controlled the Aeroflow ventilation by adjustable valves housed in the face-level vents. **M**

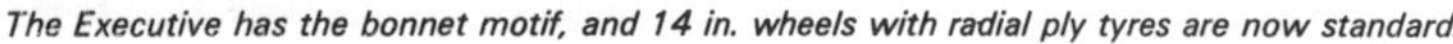
The Executive has the bonnet motif, and 14 in. wheels with radial ply tyres are now standard.

Ventilated reclining seats are a feature of the Zephyr V6 de luxe.

CROSS-FLOW CORTINA

Continued from page 36

The gearlever is also well-placed and works a beautifully smooth gearbox, although the throws are a little longer fore and aft. Another modification is that reverse is through a gate towards the driver and back, instead of the lift, twist-wrist and shove it was before.

Rear seat legroom is not altogether marvellous, and kindly people in the front seat would go forward a notch or two for the sake of comfort for those in the back. One consolation is that the smallish petrol tank (10 gallons) would probably mean a fuel stop every 220 miles or so, so there would be lots of chances to stretch the legs.

There is no problem with boot room, as the car actually has about as much space as an EH Holden — certainly the best four cylinder on sale at the moment. There is also good room to work around the engine, dominated as it is by the twin-choke Weber carburettor which still has the most infernally tricky air cleaner in the business to remove. One always drops the mounting screws and washers (there are four of them) down the carburettor throats.

And so you have it — a Cortina GT restored to something like its former glory. At $2530 it has to compete against its own brother Falcon and the Holden on power and size, but, will of course relatively beat the ears off them in integrity, roadability and handling. You can't win races with them any more, but they still have a tremendous amount to recommend them. #

THE FORD CORTINA 1600E

THE Ford Cortina 1600E, new at the Motor Show, can be described as the latest Cortina with the new cross-flow bowl-in-piston Weber-carburetted GT 1,599 c.c. 93 (gross) b.h.p. engine and the lowered and stiffened Cortina-Lotus suspension, to which luxury embellishments have been added. The last-named include black-painted radiator grille, and a painted colour strip along the side of the 4-door body, black pile carpeting, extra sound insulation, a padded arm-rest to that useful between-the-front-seats stowage bin, shaped, fully-reclining bucket front seats, a glossy wood facia and deep matching wood door-cappings, an aluminium-spoked steering wheel with leather gaiter, gaitered gear lever, dual-tone horns, cigar-lighter, spare wheel cover, "E"-motifs on the body panels, twin Wipac stainless 562 spot-lamps, to supplement the Lucas lighting, automatic very bright reversing lamps and those Rubery Owen Rostyle pressed-steel wheels, chromium-plated, which look like magnesium-alloy rally wheels or trendy U.S.A. styling to the uninitiated. These " pretty " wheels are, however, of 5½J size, shod with radial-ply tyres, so road-holding benefits.

I have had considerable experience of both the pre-1968 Cortina GT and the latest Cortina-Lotus and am very enthusiastic about both these Fords. The 1600E costs about £100 more than the 4-door GT but is £90 less expensive than the Cortina-Lotus with its twin-cam engine. quoting pre-devaluation prices.

My feeling is that, while there is certainly a demand for luxury family cars—we have in the past praised the Princess 1100 and Triumph 1300 for filling this role admirably in the smaller categories —I do not think a wooden interior decor looks right in a modern " tin " saloon, particularly, as in the case of the 1600E, it is blended with PVC upholstery, " racy " wheels, and a general impression of high-performance. Indeed, I object to this treatment, even though the veneers match well in this Ford (although I do not suppose Rolls-Royce would approve of the visible Allan holding-down screws!) on two counts: (1) that this styling is out-of-place on the smaller family cars unless used with leather upholstery and other true-tokens of sumptuousness and, (2) that its use on such vehicles diminishes the presige such decor should have when applied to large sedate, expensive cars.

Be that as it may, I expect Ford Market Research has satisfied itself that there are sufficient self-styled executives about to merit the introduction of an E-model even in the under-£1,000 price bracket. So let's take a quick look at this Cortina 1600E.

The first thing I noticed were small but subtle changes from the older GT I used to know so well. For instance, the Styla steering wheel is thick-rimmed, although not so obviously sponge-padded as the wheel of the Cortina-Lotus, but it is small and well out of the driver's line of vision. The choke can be locked fully-out but it no longer twists to hold it in intermediate positions, which could be done on the older car. The bigger gear lever knob is not entirely likeable. The wooden facia has necessitated recessing the flick switches, which is probably a good safety move; maybe it is just imagination that this has rendered them more " fumbly." The ignition-key now has a guard to prevent damage to the flywheel ring when the engine is running. The unlockable cubby-hole lid in the new wooden facia is too small to take objects which would go into the older cubby space; there is a shelf under the facia on the nearside. The padded T-piece of the pull-out-and-twist under-facia handbrake feels a bit clumsy. But the change I really object to is that the facia-vents of the ever-excellent Ford " Aeroflow " ventilation system now have their volume controls as twist-knobs in the vent centres, so it is almost impossible for the driver to reach the nearside one, which is unfortunate if he wants to adjust it while alone in the car; the former pull-out knob adjustments on the facia were much more convenient. The Kienzle clock low down on the console is illuminated when the instrument lighting is on, but the heater controls remain in darkness. At times there was a noise below the facia, emanating presumably from the speedometer drive and some minor rattles.

These aspects of the new Ford apart, it is, like all Cortinas, a thoroughly commendable, very fast and accelerative family car of capacious carrying capacity, as I have observed previously of these dependable best-selling cars. Admittedly the suspension is not outstanding and the road-holding only fully acceptable with wide-base wheels and the right tyres. The test car was on India Autobands, whereas Fords usually come to us on Goodyears or Pirellis. I did no high-speed driving in the 1600E, so could form no proper opinion of these Indias, although they have transformed the road-clinging of the children's ancient VW and the Ford seemed to have adequate traction on snowy gradients. In dry road conditions, however, another driver found the road-clinging sufficiently reassuring to make him stop the car, get out and look to see what kind of tyres it was running on. The brakes of the 1600E are adequate rather than memorably powerful.

The clutch pedal is very light to depress, if not quite so noticeably light as that of the Cortina-Lotus, which makes it insensitive. As there is a good deal of free play in the transmission, low speed running tends to become jerky. Generally, however, there is a better sense of " feel " to all the controls than on earlier Cortinas. The gearbox remains one of the best in the family-car field, but its synchromesh could just be beaten when snatching a change from 2nd to 3rd gear and the ratios are still not entirely suited to the car. The gearing is such that if the speed falls much below the built-up area legal limit it is desirable to drop into third, yet at the Castle-limit maximum the engine, then doing around 4,000 r.p.m., becomes noisy.

The first snow of winter coinciding with the test of the 1600E, I did not take any performance figures, but this is clearly a very high-performance family saloon. After standing in the open covered in ice it fired first turn of the ignition-key and only two turns were required to start it after it had remained for a day and a night buried under the snow. The Weber change-over flat-spot is faintly discernible when accelerating. Consumption of premium petrol came out at a useful 28.2 m.p.g., driving fairly normally, and from the gauge-reading the range from full to fill-up-needed would be taken by most owners as around 240 miles. The driving seat is comfortable, and the ratchet-type reclining squab adjusts over quite a fine range.

As usual, the Ford's heater provided plenty of warmth but on the test car it roasted the driver's feet while the front-seat passenger was receiving only luke-warm air. The usual row of elevated, hooded dials record water temperature, generator-charge, oil-pressure and fuel contents, the tachometer markings indicate that the engine can be taken to 6,000 r.p.m., and there is now an internal bonnet-release (the bonnet still has to be propped-up), while the sceen is of wide-zone toughened Triplex. There is a clear decimal reading on the total odometer, but no trip recorder.

The dipstick which, like the small Ford Kwik-Fill battery, is very accessible, showed that after 720 miles no oil was required. The price at time of test was £982 2s. 1d. The Cortina-Lotus, with its extra 24 b.h.p. gives even more performance more smoothly, but for those who want a de luxe GT Cortina the 1600E will fill the bill admirably especially as the new car has notably good acceleration—W.B.

FORD CORTINA + ZODIAC ENGINE DOESN'T MAKE A

SAVAGE

BY MARTIN LEWIS

Photography: MIKE COOPER

"YES, it's a standard Ford," we told the man at the pumps. "Standard Cortina GT body, standard Zodiac engine. . ." The owner of a nearby "real" Cortina did a splendid double-take before we stopped the engine. He knew that in-line fours do not make that V6 burble-burble. He could not believe his ears. When we opened the bonnet to check the oil he could contain himself no longer and almost ran across to have a look inside.

In fact, the Race Proved Savage is far from being savage. People in the office have suggested that it should be called the Model T (for torque) or the Ford Pussyfoot. However, if the driver chooses to behave like a savage, the car will join the tribe and give him all the wild thrills he can take. The idea of putting a big V6 engine in a Cortina or Corsair bodyshell is not new. At Ford's Competition Department they have been doing this for some time with estate car Corsairs used for rally service.

It is not just a matter of hauling out the 1·5-litre engine and plonking in a 3-litre Zodiac unit. Geoff Uren, the boss man at Race Proved, has spent many long hours sorting the suspension and braking to make this car as safe as it is. The Savage we tested started life as a GT, although future production will be based on Lotus versions.

When the standard engine is removed, the front cross member has to be modified to fit under the deeper sump of the V6 unit; new engine mountings have to be welded in, and an adaptor plate fitted between the clutch and gearbox, which is the standard GT and Lotus variety. The Corsair casing is a direct fit on to the V6 engine, but will not fit into the body of a Cortina. At the front a larger radiator, with electric cooling fan, is installed, while at the rear a Powr-Lok differential and beefed-up half-shafts make sure that the power is put where it is needed. Road holding is further improved

Left: Limited-slip differential and radial-ply tyres are standard. Above: Any type of special Contour seat can be chosen by the customer. A leather-covered steering wheel is fitted and rev counter is taken from the Ford Zodiac

by wide-rim (5·5 in.) wheels shod with Goodyear G800s.

The result is about as near to an American car as one can get in Britain—a light body backed up with a great deal of power and torque from a relatively lazy engine. The V6 Zodiac engine develops 136 b.h.p. net at only 4,750 r.p.m., and a mighty 192 lb. ft. of torque at 3,000 r.p.m. The rev counter is red-lined at 6,000 r.p.m., although we did sneak another 500 r.p.m. when trying for ultimate performance. According to the gearing, however, the rev counter must have been over-reading by about 10 per cent, and a true 6,500 r.p.m. in top (which the car obviously will do) means 116 m.p.h.

With just a dab on the accelerator pedal to set the automatic choke and a twist of the ignition key, the Savage rumbles into life. The V6 throb is the essence of this car, accentuated by the beat from the carburettor air intake trunking and the noise from the two exhaust manifolds which run into a single pipe under the car. Never does the car make an obtrusive noise, but the gear lever did chatter above 4,500 r.p.m.

This is very much a two-faced savage. For the morning commute, it can be driven in two gears—second and top. Even on the open road top and third are the only gears really used, the torque making it unnecessary to change down farther. New drivers tend to go zipping up and down through the box in true Cortina fashion, and the temptation to change down when overtaking is hard to overcome, when all that is needed is a little more loud pedal.

Although the car is so docile and well-behaved in town, it really does hitch up its skirts and get moving on the open road, in a manner that will leave practically everything else standing. The standard Lotus final drive ratio of 3·78 to 1 is rather too low, allowing the engine to run out of breath too soon; the maximum speeds in the gears at 6,000 r.p.m. are 33, 50 and 70 m.p.h. We tried a variety of ways for the standing start acceleration figures, and found that about 3,000 r.p.m. produced just the right amount of wheelspin before the tyres dug in and powered the Savage away. Changing at 6,000 r.p.m. produced some pretty good times, but 1,000 r.p.m. fewer kept the engine far nearer the top of its power curve, and gave even better results. This sort of drag take-off is motoring indeed, and not until one gets into the realms of the 4·7-litre Ford Mustang, 7-litre Olds Toronado or Jaguar 420 can one really start to make comparisons. The Savage is even faster than the ultra-low geared Group 2 works Cortina-Lotus which we are reporting on soon.

In terms of bare figures, the Savage is 1·5sec faster to 50 m.p.h. than the Cortina-Lotus and takes 2·8sec fewer than its "parent" Executive Zodiac, with automatic transmission. The 70 m.p.h. mark is reached in 12·2sec—14·9sec for the Lotus and 17·9sec for the Zodiac. The standing quarter mile is covered in a mere 16·6sec—with the Lotus 1·6sec behind and the Zodiac a full 2sec slower.

The Savage's top speed is above the 100 m.p.h. mark with this rather low gearing and on Britain's restricted roads, 70 m.p.h. is equivalent to around 4,000 r.p.m. The low final drive ratio also accounts for the rather poor fuel consumption of 18·9 m.p.g. However, the enthusiastic driving of four staff members in the few days we had the car did not exactly make for economy. With a higher axle ratio and less "boot," the claimed 28 m.p.g. might be within reach. Part of the Race Proved package includes an additional 8gal tank located in the front of the boot, so that a total of 18 gal can be carried. As the fuel gauge is in the lower tank, it does not start moving from the full mark until the top tank has been emptied.

We did have a little trouble with the car boiling briefly when the engine was switched off after a long, hard drive. Geoff Uren says that the thermostat for the electric fan is at present mounted on top of the main cooling thermostat housing and this layout does not work too well; on future models this arrangement will probably be altered. With its big fat tyres, adjustable dampers, double anti-roll bars, and carefully developed braking and spring rates, the Savage's handling is impeccable. Despite the firmer-than-usual ride, the wheels stay firmly in contact with the ground, and the power can be used easily to help the car round corners. Wet roads have to be treated with a certain amount of caution, as an injudicious right foot can cause a bit of snaking. Braking is well up to the car's performance, although there is quite a lot of pedal travel before the linings start to bite.

Inside, the Savage is virtually identical with the standard GT or Lotus: The Zodiac rev counter is used, as it matches the engine, and any combination of Contour seats can be fitted at the front. We had the 6 for the driver (fixed back, deep side rolls) and the 7 (reclining back, shallow side rolls) for the passenger.

It is not often that the test staff are so unanimous in their praise for a car. To use the Savage seems the ideal form of modern transport—very fast off the mark, compact enough to be used in towns, yet large enough to pack in the family and luggage and go rushing off to the South of France. The smoothness with which the power is delivered makes it incredibly easy and restful to drive; unlike those of so many "tuned" cars, the Savage engine rarely needs to rev above 4,000, and should give years of life. And, of course, it is all standard Ford. ■

PRICE

Complete new car with purchase tax paid .... £1,365 From Race Proved Performance and Racing Equipment Ltd., 117 Uxbridge Road, Hanwell, London, W.7. Telephone 01-579 0991.

Figures in brackets are for the Ford Cortina-Lotus tested in Autocar *of 17 August, 1967.*

Acceleration times (mean): *Speed range, gear ratios and time in seconds:*

m.p.h.	Top (3·78)	3rd (5·28)	2nd (8·86)	1st (11·21)
10-30	6·3 (10·0)	4·6 (7·7)	3·0 (4·4)	2·1 (3·0)
20-40	5·8 (10·0)	4·0 (6·7)	2·8 (4·1)	—
30-50	6·0 (9·8)	3·8 (6·3)	3·41 (4·5)	—
40-60	5·8 (10·6)	4·2 (6·4)	—	—
50-70	5·5 (12·4)	5.6 (6·9)	—	—
60-80	6·7 (15·4)	— (8·8)	—	—
70-90	9·2 (18·7)	—	—	—
80-100	11·7 (23·2)	—	—	—

From rest through gears to:

30 m.p.h.	2·7 sec	(3·6 sec)
40 „	4·5 „	(5·6 „)
50 „	6·4 „	(7·9 „)
60 „	8·8 „	(11·0 „)
70 „	12·2 „	(14·9 „)
80 „	16·0 „	(20·1 „)
90 „	21·3 „	(30·9 „)
100 „	36·3 „	(44·0 „)

Standing quarter-mile 16·6 sec (18·2 sec)
Standing kilometre 31·1 sec (32·2 sec)
Maximum speeds in gears

	m.p.h.	k.p.h.
Top (mean)	**104** (104)	**167** (167)
(best)	**104** (105)	**167** (169)
3rd	**70** (83)	**113** (138)
2nd	**50** (58)	**80** (93)
1st	**33** (39)	**53** (63)

Overall fuel consumption for 461 miles:
18·9 m.p.g.; 15·0 litres/100km
(22·2 m.p.g.; 12·7 litres/100km.)

Whether coming or going, the latest Cortina has crisp workmanlike lines uncluttered by chrome styling gimmicks.

12,000 mile staff car report

Ford Cortina GT

'It...seemed very appropriate...'

by Philip Turner

Photographs by Maurice Rowe

"DOES your office *always* look like this?" an awed visitor asked me recently as he gazed around on the piles of motoring magazines in umpteen languages that surrounded me on three sides. Occasionally, there is a rumbling thud followed at once by terrible Turner oaths which signifies merely that another over-tall pile has leaned rather more than the Tower of Pisa and subsided in chaos on the floor. In addition to *Quattroruote* from Italy, *Automovilismo* from Spain, *Car* from South Africa, *Australian Motor, L'Action Automobile* from France and many, many others, I also study the club magazines with very great interest, for they often provide the best indication of all the new trends in the sporting world. In one club magazine recently, the editor, in his chat column, remarked that yet another member of the committee had recently acquired a Cortina G T, so that now practically the entire committee consisted of Cortina G T owners.

It is true that motor club members in Britain today are usually either young 'uns in Minis or family men in Cortina GTs. It therefore seemed very appropriate that *Motor's* sports editor's next car should be the latest of the Cortina GT breed at that time – Spring, 1967 – i.e. the new body shell but with the old 1,500 c.c. engine. After taking advice from Cortina experts, I asked for Pirelli Cinturato tyres on the wider $4\frac{1}{2}$ rims, and was also fortunate in that, by the time I received my car, the revised gearbox with the higher second gear to eliminate the over-large gap between second and third speeds was being fitted as standard. I was undecided whether or not to have the two door or the four door model. I preferred the look of the two-door body, but thought that the four-door one would be more practical. When delivered to me, on March 28, 1967, the Cortina was in the specified dark green finish with black upholstery and was, in fact, the four-door version. It was at once equipped with outside mirrors and seat belts, and the running-in began.

It must be admitted that first impressions were not particularly favourable. After the superb comfort of the Renault 16 suspension and seats, the Cortina felt like a 10-year retreat into the past. Wind noise, too, seemed lamentably high and any inequalities in the road surface produced loud thuds from underneath. I took delivery on a Tuesday. A run to Coventry and back on Friday, eschewing M1 in favour of A41/A423 on the outward run and A5 on the return as being much less frustrating when running in, followed by a visit to the lions at Longleat on Saturday saw 500 miles completed by the weekend and the car received its first, 500-mile, service on the following Wednesday.

By this time, the rev-counter has ceased to function, a thin film of oil had spread its way along the left hand parcel shelf and the rear seat cushion inconvenienced rear seat passengers by persistently sliding forward. The oil leak was traced to a loose connection to the oil gauge and was duly dealt with, but no replacement rev counters were available and the rear seat cushion continued its advance.

The next weekend the Ford went abroad for the first time to

convey us to Le Mans to cover the Test Days, during which the first 1,000 miles appeared on the odometer and, as usual, we explored all those somewhat rugged tracks that lead to Mulsanne and Arnage on the inside of the circuit. As on the following weekends the Ford and I went to Oulton Park for the Spring Cup meeting, to Silverstone for the Vintage meeting and to Silverstone again for the International Trophy, the Cortina had covered over 2,400 miles when it returned to Highbury Motors for the new rev-counter at last to be fitted. From May 3 to 19, however, the car lay fallow in my garage while I reported the Monaco GP then drove to Sicily and back in a Fiat 124 Sport coupé to cover the Targa Florio. And from May 24 to May 30 it again lay idle while I was in Greece reporting the Acropolis Rally in a works Mini. However, in June the Cortina began to pile up the miles again, taking me to Zandvoort for the Dutch GP to watch the Ford GP engine win its first race, then on to Le Mans to report the 24 hours. And in all the times I have been to Le Mans, this was the first occasion on which I drove away afterwards in a car of the winning make.

After a precious weekend at home – well, in England at any rate – the next two weekends involved out and home rush to Rheims for the 12 Hours and to Le Mans for that pathetic French Grand Prix. While at Le Mans, I took the Cortina into the garage at which the Ford Le Mans team was based in 1965 for a sticking throttle to receive attention. Some oil expertly directed into the right place – the pivot on the front bulkhead – did the trick and the car no longer motored on at full chat when I lifted off, a habit that ensured life was never dull.

By now the Cortina had covered nearly 6,000 miles, and quite suddenly, I began to like the car very much indeed. Suddenly, it had come to life and from being a rather dull and uninteresting machine it had developed a truly sparkling performance. I have since learnt that this is common practice for the G T, that for the first 5,000 miles or so it is apt to prove rather disappointing, but then almost overnight it loosens up and really begins to travel. Ever since, the Cortina's performance has been a constant delight, for I do like a lively car.

Other plus points, too, were becoming more obvious. The gearchange is light, very quick and a pleasure to use. Which is important for an old fashioned driver like me who lives on the gearbox, rather than the brakes. Moreover, all the ratios in the box are very usable. Bottom is quite high so that one can accelerate smartly away from the line in it, then up into second which usually suffices to take me well clear of the opposition should I feel so inclined. Personally, I long ago came to the conclusion that to treat every traffic light as a starting grid is a waste of nervous energy, quite apart from shooting one out into the path of some foolish character hurtling across on the red. But it's nice to know that you can if you want to, and there are traffic situations which make it preferable to draw ahead rather than to continue accelerating side by side.

Continued overleaf

At work—the Cortina parked behind the Press stand at Le Mans last year. Intense looking gent on the right is the editor about to deposit brief case and other chattels.

12,000-mile staff car report

The steering was very much better than I had expected from previous comments that talked of springiness in the system. To me, it seemed pleasantly precise, and although quite light one could feel what the front wheels were up to. Running the Pirelli Cinturatos at 28 lbs. all round as recommended to me by experienced Cortina users resulted in a car that understeered its way round corners with a certain amount of body lean but no drama. On winding roads on a good surface the car can be hurried along without feeling at all like a family saloon being over pressed.

On less than excellent roads, however, the car was not so happy. My standard of performance in France is a reasonably driven Peugeot 404. On well-surfaced Routes Nationale the Cortina GT will sing along at normal (not 404 C fuel-injected) Peugeot 404 cruising speeds, but on the typical wavy surfaced three figure N roads and double figure D roads it cannot live with the 404s and is unhappy at much over 70 m.p.h. A suspension king, with whom I spent an hour or so drinking afternoon tea while awaiting an air ferry at Calais Airport told me he had encountered just the same trouble with his Cortina GT at high speeds and had improved the car greatly by substituting Cortina Lotus type dampers at the front—Armstrong RAS 1731 on the left and RAS 1732 on the right. I have been meaning to try this mod ever since, but have yet to get around to it. Apparently, the valving on the Lotus type Armstrong dampers is much more sophisticated and is therefore better able to cope with the pattering of the front wheels on such roads.

I like the driving position very much, for I like sitting fairly upright and looking out on the world through a big, deep windscreen. The fact that the rear seat squab becomes more vertical as one forces the seat rearwards on its runners does not trouble me but the effort needed to alter the position of the seat and the spiteful habit the operating lever has of biting bits out of one's fingers is not merely troublesome, it is exasperating. On the other hand, if the designers had to save a few pennies in any details of the car, I would rather they saved them on the seat mechanism than on the engine or suspension.

The seat itself has proved comfortable even when it has been occupied all day. No complaints about the pedals either, which are so laid out that one can heel and toe with ease. My chief complaint about the switches is that, as on most British cars, one has to fiddle about on the instrument panel to find the switch to operate the lighting system. That the lights should not all be controlled Continental fashion by a single lever projecting from the steering column is irritating enough, but to add to my ill temper the lighting switch is located behind the steering wheel by which it is obstructed.

In town, the outstandingly good steering lock which provides a turning circle of only 25 feet is a boon and a blessing; even Turner can park the Cortina in a gap that seems impossibly small at first sight. Designers who produce cars for today's crowded conditions with five acre field turning circles should be doomed for evermore to drive the streets of London striving to park in gaps that are adequate for less clumsy cars.

As a result of all this dashing about, the Cortina had covered over 6,000 miles before it received its 5,000 mile service, when the rear seat squab was at last made to stay put. The exhaust system was also attended to, for the dull thuds from underneath had grown duller and thuddier, and I noted that a case travelling in the boot above the exhaust pipe was done to a turn by a hot spot in the floor by journey's end.

The Cortina stayed in England for the rest of July, but in the first week in August set off for foreign parts again, first to Germany for the German Grand Prix then northwards up the autobahn system and to Denmark via the Puttgarden-Rodby ferry. Joining the autobahn before Bonn, we pressed steadily on

Continued on the next page

Waiting to board the good ship Winston Churchill at Esbjerg harbour. Turner junior on the right stretches his legs.

Performance

(No comparisons possible as GT version of Cortina not road tested in current form)

Maximum speed (mean), 91.0 m.p.h. (best), 94.74 m.p.h.

Acceleration in upper ratios	**Top**	**3rd**
20-40 m.p.h.	9.9	6.2
30-50 m.p.h.	9.4	6.2
40-60 m.p.h.	10.4	7.0
50-70 m.p.h.	11.5	9.2
60-80 m.p.h.	14.9	—

Standing start acceleration figures	
0-30 m.p.h.	4.1
0-40 m.p.h.	6.3
0-50 m.p.h.	9.4
0-60 m.p.h.	13.5
0-70 m.p.h.	18.8
0-80 m.p.h.	27.9
Standing quarter mile	19.2

Fuel consumption	
at 30 m.p.h.	44.0 m.p.g.
at 40 m.p.h.	42.6 m.p.g.
at 50 m.p.h.	36.9 m.p.g.
at 60 m.p.h.	32.1 m.p.g.
at 70 m.p.h.	27.0 m.p.g.
at 80 m.p.h.	21.6 m.p.g.
at 90 m.p.h.	17.1 m.p.g.
Overall	25.8 m.p.g.

Speedometer error						
Indicated	30	40	50	61	72	83
True	30	40	50	60	70	80

Distance recorder 1.8% fast

What it cost

Running costs	£	s.	d.
Road Fund licence	17	10	0
Service at 5,000 miles	5	17	4
Service at 10,000 miles	5	5	6
Tyre wear (40% on four tyres)	12	2	2
New sparking plugs		12	0
New air filter element		8	6
Petrol: 494 gallons at 5s. 8d. a gallon	139	19	4
Oil: 25½ pints	3	16	6
Running costs, total	185	11	4
Insurance, AOA Group 4*	60	0	0
Price of car new	880	0	0
Second hand value (approximately)	660	0	0
Cost/mile (excluding depreciation and insurance)			3.71d.
Cost/mile (including depreciation and insurance)			9.31d.

*A hypothetical case based on comprehensive cover for Group 4 (AOA rating) in Metropolitan (district 3) area without no-claim discount.

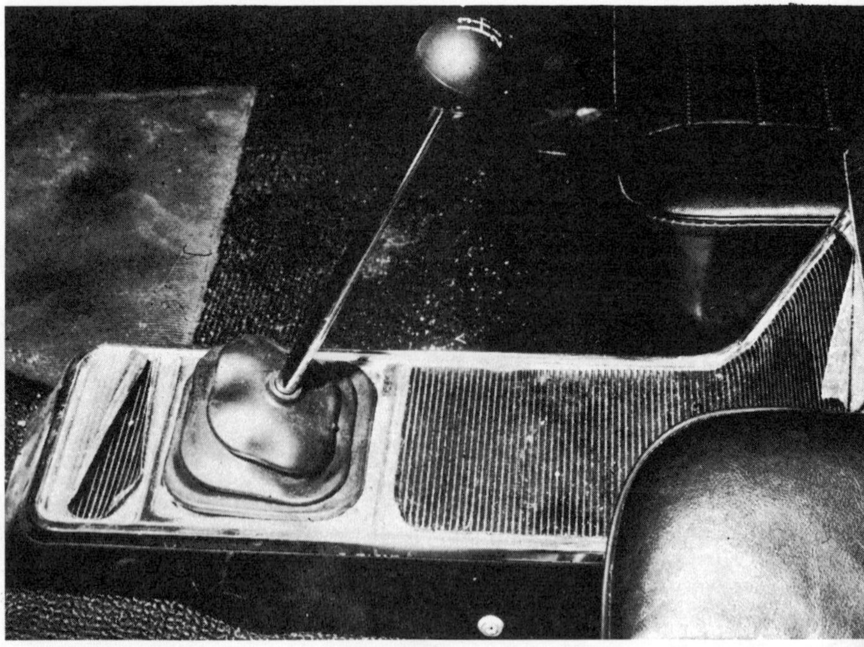

Covering of the central console peeling off from both ends—at the 5,000-mile services the edges were stuck down and have stayed that way ever since.

12,000-mile staff car report
continued

via Cologne, where it is all too easy to take the branch to the west of that city instead of the one to the east, then to Hanover, the outskirts of Hamburg and to Lubeck, where we stopped for the night.

This 400 miles plus of continuous autobahn motoring brought home to me very forcibly that while England is the country where one increases speed to 70 m.p.h., on the Continent one slows to 70 m.p.h. Cruising speeds on German motorways especially have increased considerably, so that now the bigger Fords and Opels as well as the BMWs, Mercedes and Porsches all cruise at around 90 m.p.h. Moreover, driving manners on the autobahnen have improved considerably and there is far less baulking than there used to be. The Cortina swept steadily on with its cruising speed rising from around 80 m.p.h. to just on 90 m.p.h. indicated, equivalent to a genuine 87 m.p.h.

It is when one holds such speeds almost by the hour that the sheer robustness of the Cortina becomes ever more appreciated. I have for long thought that motoring journalists require cars as ruggedly built as kitchen alarm clocks rather than the exotics made like Swiss watches. The Cortina is a very fine car indeed for anyone who requires a car for hard driving, rather than for polishing on Sunday mornings outside his garden gate.

In the small hotel in Lubeck where we stayed the night there was a mystery, for hanging on the bedroom wall was a painting with a large bullet hole in it. Was it the war, when the RAF did great damage to this old city, a crime of passion or a pistol packing visitor who did not like the picture? Alas, we never found out.

For the next fortnight or so the Cortina found itself being used as a family car for a change, for after a few days in Copenhagen—a splendid city but with most expensive restaurants—we drove up the coast to Gilleleje, there to sit on the beach, swim and watch the splendid fleet of fishing boats sailing in and out of the harbour. Sailing is perhaps not quite the word, as they were propelled by the slowest revving diesel engines I've ever encountered.

The ability of the boot to swallow vast quantities of luggage is most useful on a family type holiday, and the lack of polished wood and other trimmings for the interior is a definite asset on such occasions.

An advantage of running a Ford is that wherever one goes there always seem to be Ford dealers around. Not that the Cortina

Continued on the next page

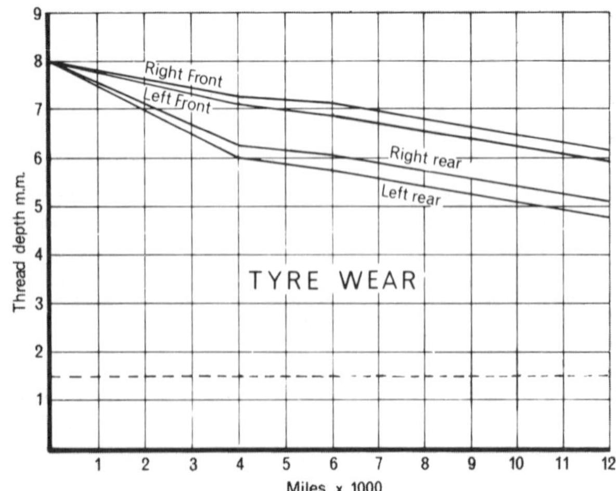

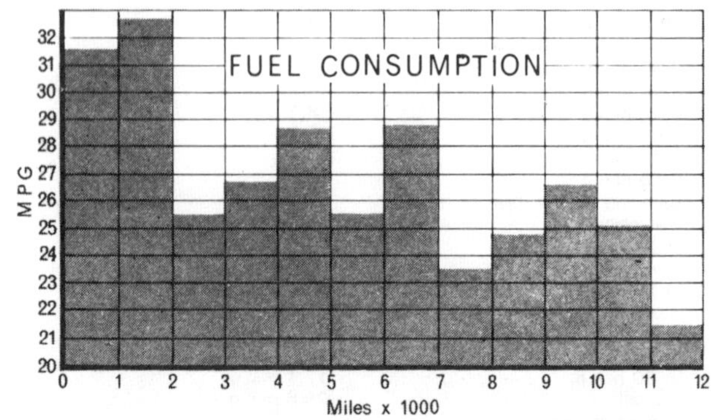

12,000-mile staff car report
continued

required the slightest attention in Denmark, but it is comforting to know that help is at hand should it be required.

By the time the Ford returned to England aboard the *m.v. Winston Churchill,* it had covered exactly 1,570 miles since setting off for the Continent. For the rest of the year it remained in Britain. At the 10,000 mile service in September, which happened when 10,443 miles had been covered, the only fault requiring attention was the obstinate refusal of the driver's door to close properly, unless precisely the right pull was exerted. Otherwise, it would appear to be shut but in fact would be on the second catch.

As, however, the door did not rattle when in this position—the whole body seems very rigid and free from rattles—I only found that it was not shutting correctly by the increase in wind noise. An attempt had been made to cure this fault at the 5,000 mile service, but without success. On this occasion, however, the cure was complete, and the door would gently pull to the shut position without any need to slam it. Very useful when returning late at night from a dinner for one can then shut and lock all doors without arousing the entire neighbourhood. A glum friend of mine with a flat in town once swore there were no two-door or four-door saloons after 1 a.m.—they were all 14 door saloons.

A Crypton check at 12,966 miles showed that a set of new plugs was required, the air filter element needed replacing and the tappets required adjustment. When these matters were attended to, an oil leak by the oil pump was also cured, thereby restoring the consumption of oil to its old figure of around 1,000 miles a pint.

One of our readers who is also a Cortina GT owner sent an accompanying letter when he returned the completed questionnaire urging me to write fearlessly and honestly about the Cortina and to pull no punches. Not that I generally do. . . . But in this case, why should I, for there are precious few punches to deliver, let alone pull. The car has always started without any hesitation whether left overnight in a garage or standing outside some hotel in France, Belgium, Germany, Holland or Denmark. It has never been halted on the road with any trouble. It has never objected to being driven very quickly for hours on end. It has never displayed any bad habits when I rushed into a corner with an excess of enthusiasm over judgment. What then have I to complain about?

Admittedly, it is not the most refined car in the world, but then as one reader wrote in reply to the question. Would you buy another of this model or make? "Can you suggest a better car for £900. I'm listening." Wind noise at speed is perhaps its major fault, for the constant rushing noise can be tiring on long, fast runs. The Ford decision to revert to one-piece doors for the Escort is most interesting in this connection, for I am sure that fabricated half-frame doors are the chief culprits. One day, when I have time on my hands, I want to try the Cortina with the doors sealed with Sellotape. Not permanently, of course. **M**

Second opinion

Philip Turner's Cortina GT nearly shattered my previous impressions that the GT is an ideal sporting family carriage; this had been based very much on my own Cortina 1500 of the previous style. I didn't like its replacement so much—its handling had gone all soggy—but the GT had preserved the same spring rates and should have felt much the same. Philip's car, however, felt pretty well worn when I first tried it; every bump in the road created its own noise transmitted particularly through the rear radius arms, and the tyre pressures felt far too high—dropping them from Philip's 28 to the recommended touring 24 p.s.i. brings the low speed radial thump to an acceptable level. These upper radius arms fitted to the GTs create a very rigid drive line which calls for practised clutch work for smooth thump-free changes. The steering and handling felt wallowy at the front end, which I checked by bouncing on the car later and found that the front dampers are hardly working as the nose completes about two cycles of movement after you stop bouncing. Those at the rear are probably on the way out, too, but do have some life left in them; damper life is, of course, considerably shortened by the Continental road surfaces which constitute a large part of the sports editor's mileage—export settings are probably firmer.

Having run my own two-door Cortina and driven Philip's four-door one I think I would stick to two doors as the four-door does not feel as rigid, although this impression has subsided a little since the tyre pressures have been lowered.

I thought the wind noise was totally unacceptable and I could not have lived with it as long as Philip has done.

Once I had accepted that the car was suffering only from two curable faults—dampers and wind noise—I enjoyed it considerably more. Its acceleration is usefully lively although it didn't feel quite as fast as I remembered and the gearbox feels just as delightful as it should on a *new* car; the previous version became very imprecise after this sort of mileage. Roadholding is good, particularly in the wet, and the Cinturatos have a lot of life in them—enough for nearly 24,000 miles. In general this car has survived its first 12,000 miles of hard motoring quite well, paintwork and interior being good, but its dampers should last better and there is little excuse for bad wind noise with fixed quarter lights.

M.H.L.B.

Comment from Ford

Considering the performance and reliability of Mr. Turner's Cortina GT there is very little constructive comment that we can make. His model, of course, has now been superseded by the 1600 c.c. cross-flow Cortina GT which is giving improved performance and fuel economy.

Latest production Cortinas are now fitted with an improved door seal and this goes a long way to reduce wind noise.

The recommended tyre pressure for 165 radials on the 1,498 c.c. Cortina GT is 24 lb. per square inch at front and rear and in our experience the increase to 28 lb. per square inch will vastly reduce the ride quality without a noticeable improvement in roadholding.

Other owners' comments

Of the 30 readers who kindly completed our questionnaire forms —and may I say here how much we appreciate the co-operation we receive—20 said they would buy another Ford Cortina GT. Seven would not and three were undecided. Owners of early production models of what perhaps we may call the Mk. 2 Cortina, to distinguish it from its predecessors, had the most trouble and were the least likely to buy another one. Judging from the experience of our readers, these early cars were prone to gearbox trouble, especially jumping out of second, and they also suffered a certain amount of clutch trouble. The brakes, too, which in many later questionnaires were dubbed either satisfactory or excellent, gave trouble to three out of four of the owners who bought their GT in November, 1966.

Praise was meted out by 13 owners for the heating and ventilation system, 10 the excellent turning circle, nine the good performance, seven commended the comfort, five the handling and four the capacious boot. The handling might have called for even more favourable comment if the 4½in. wide rims and radial tyres were fitted as standard, for those owners who had specified these when ordering their cars were well pleased with the way the GT handled. On the other hand, some owners running on the standard rims and tyres complained strongly of lack of rear wheel adhesion in the wet.

Fourteen owners complained bitterly of the front seat adjustment; they did not like the mechanism and did not approve of the driving positions it provided. Wind roar at speed irked 10 owners, and four thought the lights were inadequate for the speed. The umbrella style handbrake was disliked by six.

Service for the car was praised by 16, damned by nine and the rest were rather non-committal. Obviously, previous Cortina GTs have given good service, because a number of readers mentioned that this was their third.

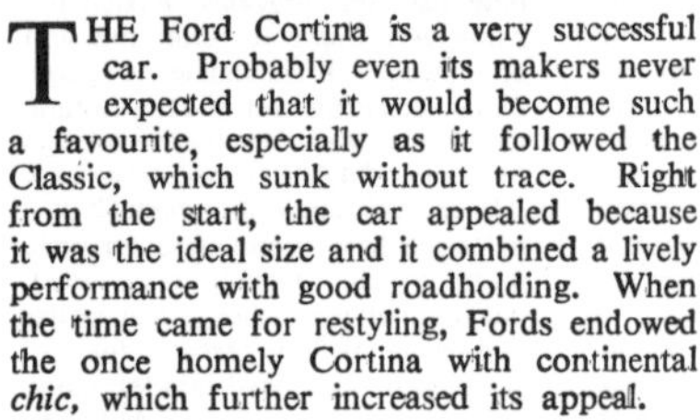

ROAD TEST
by John Bolster

FORD CORTINA 1600E

Ford's smallest executive saloon packs plenty of punch

THE Ford Cortina is a very successful car. Probably even its makers never expected that it would become such a favourite, especially as it followed the Classic, which sunk without trace. Right from the start, the car appealed because it was the ideal size and it combined a lively performance with good roadholding. When the time came for restyling, Fords endowed the once homely Cortina with continental *chic,* which further increased its appeal.

To match the new appearance, the performance has now been increased by the adoption of a crossflow engine. In the case of the 1300, this has a pure Heron head with a flat surface, the combustion chambers being recessed into the piston tops. The previous 1500 cc engine has been replaced by a 1600 cc unit, which has now become a more popular size, and in this case the bore remains the same at 80.98 mm, but there is a taller cylinder block to accommodate the extra stroke of 77.62 mm. A bowl-in-piston combustion chamber is again used, but the head is not completely flat, having shallow combustion chambers in which the valves seat, permitting the deletion of the "spectacles" for valve head clearance which are machined in the piston tops of the 1300.

The cylinder head, in both versions, now has the inlet and exhaust manifolds on opposite sides. A considerable increase in power output is given, but it is the improved torque for acceleration in the middle ranges which is the main feature, most valuable for quick overtaking without exceeding the speed limit.

The larger engine benefits from the new design and it also gains more torque from its extra 100 cc compared with its predecessor. The GT version, which is the subject of the present test, has a special camshaft and a Weber twin-choke downdraught carburetter of the compound type. This gives it an extra 10 bhp (net) over last year's model but the greater efficiency of the design is also reflected in better fuel consumption figures at all speeds.

Perhaps the most important mechanical improvement is the restaging of the gearbox ratios. The previous GT model was cursed with an extremely low second gear, and this has been rectified, both first and second speeds now having considerably higher maxima. This not only improves the performance but makes smooth driving much easier, now that the wide gap between third and second gears has been reduced. The well-tried chassis design is naturally retained, with MacPherson strut and lower wishbone front suspension and semi-elliptics behind, plus trailing links. The suspension is lowered as in the Lotus Cortina.

The car chosen for our test was the 1600E. Fords have already entered the executive market with their V6 Zodiac and they have now very wisely given the treatment to a model which is of more convenient size for many people in England. There is a considerable demand for a car which has the luxury of a craftsman-built limousine allied with the many practical advantages of a popular mass-production chassis.

The 1600E can at once be distinguished by its highly decorative wheels. This type of wheel may look rather absurd on some cars, but it seems to suit the lines of

CONTINUED ON PAGE 33

MOTOR ROAD TEST No.8/68 ● Ford Cortina 1600E

Enjoyable extrovert

Luxury Cortina; good performance and roadholding; indifferent ride; too much noise; extras good value

AMERICAN manufacturers can easily produce 10,000 variations on a basic model without a repeat, so it is hardly surprising that those car makers in Britain that get nearest to this model proliferation have American parentage. It is perhaps the success of the bolt-on goodie trade that has inspired Ford, Rootes and Vauxhall to take the plunge into the option game on their basic themes. With a little more power and/or a little more trim a manufacturer can manipulate his model into a completely different market.

Ford started on the power theme, enlarging the original Cortina 1200 to 1500, GT and then Lotus tunes, repeating themselves in the new shape. But it was not really until the appearance of the Corsair 2000E that they tapped the market which requires more obvious evidence of the best car in the range. To the Cortina GT—a brisk five-seater saloon—Ford have added extras which two different types of clientele could only add for themselves at greater cost from outside sources. E really stands for Executive—as in the Zodiac but on a lesser scale; to appeal to these "junior" executives the 1600E boasts a woodlined interior and a facia more reminiscent of a Jaguar's, a gold line down the side, reclining seats, fancy wheels and the new, squatter look which suggests that the owner might be as get-ahead as the car undoubtedly is on performance.

On its other front, where E might stand for Enthusiast (or even Extrovert), the new squat look serves a more useful purpose than just appearance; it follows from having the Cortina Lotus suspension—firmer and lower—and 5½J steel wheels cleverly styled to look like mag alloy ones. The handling and roadholding are now extremely good, particularly so on wet and greasy roads, where the India Autobands seem to excel. In fact, from a keen driver's view point there is little to complain of in the 1600E; with the standard GT engine it goes well—maximum of 96.3 m.p.h. and 0–60 m.p.h. in 11.8s.—and has only a moderate thirst for petrol, our overall consumption of 23.1 m.p.g. being pretty fair for a hard driven 18½ cwt., five-seater saloon.

It is on the Executive front that the car is not quite up to its label. It looks attractively aggressive (or vice versa) and the interior is well styled and finished but it is not particularly restful on the open road. The ride is jerky, there is too much wind noise at all out-of-town speeds, and engine harshness sets in just on 70 m.p.h. in top, giving an initial impression of undergeared fussiness. The average enthusiast will probably forgive such deficiencies since the car is such fun to drive far and fast, but our mythical executive might well want quieter transport for his £982, even though at £92 the extra bits are a bargain. Continued on page 25

PRICE: £799 plus £183 2s. 1d. purchase tax equals £982 2s. 1d. Radio £27 4s. 6d, with tax; total as tested £1,009 6s. 7d.

The back seat gives plenty of head, knee and shoulder room for adults; there is an ashtray in the back of the transmission tunnel console.

Top: Not surprising that the 1600E has a Cortina Lotus air about it—it has the Lotus wide wheels, lowered suspension and matt black grille, plus also a pair of main beam spots.

Above: Despite their mag alloy appearance, the $5\frac{1}{2}$J section wheels are styled in steel; the 1600E comes in four-door form with a thin gold line down the side and an emblem on the rear pillar.

Left: The capacious boot took 10.9 cu. ft. of our test luggage. The complete toolkit can be seen in use—one jack with ratchet spanner, one wheelbrace.

Below: The vestigial toolkit can be clipped to the floor of the boot.

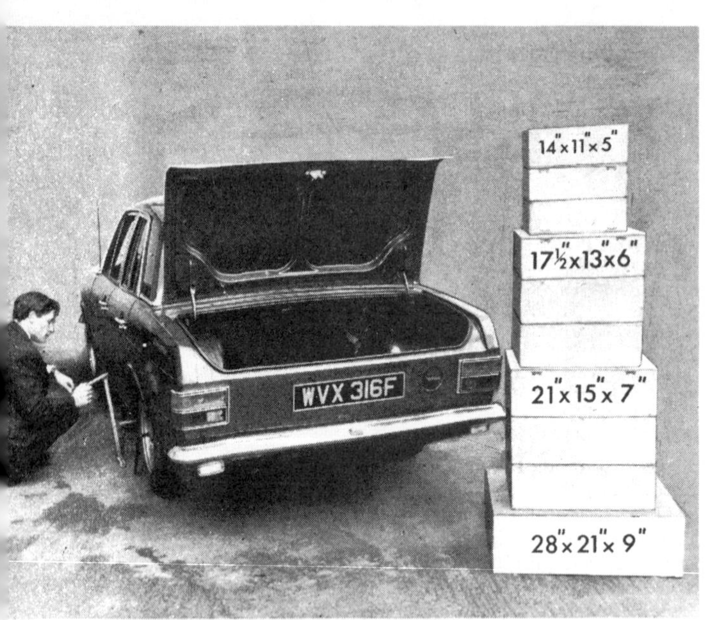

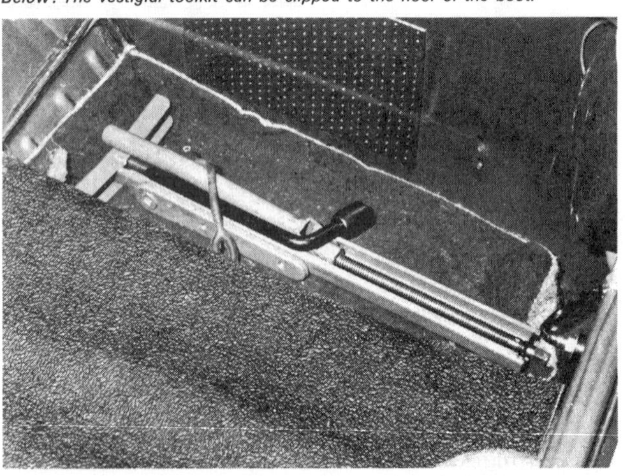

MAXIMUM SPEED

60 65 70 75 80 85 90 95 100 105 110 115 120

m.p.h.

ACCELERATION

26 24 22

seconds

Ford Cortina 1600E
£982

Fiat 125
£999

Mazda 1500
£996

Simca 1501 GLS
£1,080

Vauxhall Victor 2000
£910

Ford Cortina Lotus
£1,080

Ford Corsair 2000E
£1,039

Ford Cortina 1600E *continued*

Performance and economy

Introduced at last year's Paris Show, the 1600E uses a version of the cross-flow bowl-in-piston engine common to the rest of the range; this one is identical with that of the Cortina GT. Compared with the 1600 it has larger valves, a hotter camshaft, a twin-barrel Weber carburetter and a fabricated exhaust manifold. The output is now 88 b.h.p. from 1.6-litres against the original 78 from 1.5 of the earlier G.T.

With full choke it fired first time and pulled smoothly straight away; although it took over two miles for running temperature to appear on the gauge, the choke could be pushed in after little more than a mile. When warm it pulled well from 1,500 r.p.m. but you can't use less than this in top gear (26 m.p.h.) without getting some snatch from the harsh drive line. Above this is a usefully wide torque band and you can use all the revs to advantage, up to the indicated limit of 6,000 r.p.m. Unfortunately, a noticeably harsh period starts at about 4,000 r.p.m. which thus acts as a pottering rev limit; it is really because the engine is initially so unobtrusive that you are the more conscious of the change. This harshness is at its worst between 4,200 and 4,600 r.p.m. but still noticeable beyond although in top gear it gets drowned by wind noise at high speeds. If you are in an over-70 m.p.h. land this might get a little wearisome but the engine is certainly quite happy to maintain as much as a steady 90 m.p.h. (5,200 r.p.m.) if you want to. It is a pity really that the 3.778 final drive of the Cortina Lotus is not offered as an extra as the 3½% gearing increase would put 70 m.p.h. just below the onset of the harshness.

The gearing is just right for maximum speed at 96.3 m.p.h. and the engine pulled an indicated 6,000 r.p.m. on the fastest straight at MIRA (the rev counter being about 250 r.p.m. fast at this

Continued on the next page

Performance

Performance tests carried out by *Motor's* staff at the Motor Industry Research Association proving ground, Lindley.

Conditions

Weather: Dry but foggy, no wind.
Temperature 30°-32°F. Barometer 29.05 in. Hg.
Surface: Dry concrete and tarmacadam.
Fuel: Premium 98 octane (RM), 4-star rating.

Maximum speeds

	m.p.h.
Mean lap banked circuit	96.2
Best one-way ¼-mile	98.9
3rd gear at 6,000 r.p.m.	74
2nd gear at 6,000 r.p.m.	51½
1st gear	35

"Maximile" speed: (Timed quarter mile after 1 mile accelerating from rest)

Mean	95.8
Best	96.8

Acceleration times

m.p.h.	sec.
0-30	3.8
0-40	5.7
0-50	8.4
0-60	11.8
0-70	16.2
0-80	23.6
0-90	37.9
Standing quarter mile	18.6

m.p.h.	Top sec.	3rd sec.
10-30	—	7.2
20-40	10.6	6.6
30-50	10.2	6.3
40-60	10.1	6.5
50-70	11.5	8.3
60-80	14.2	—
70-90	22.0	—

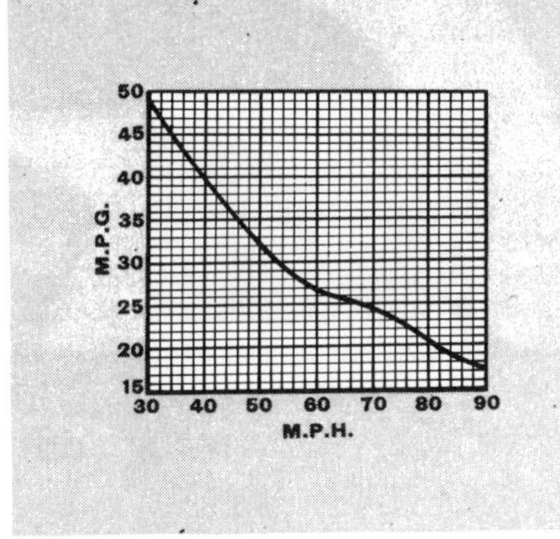

Fuel consumption

Touring (consumption midway between 30 m.p.h. and maximum less 5% allowance for acceleration) . . . 24.7 m.p.g.
Overall . . . 23.1 m.p.g.
(=12.2 litres/100 km.)
Total test figure . . . 1,150 miles
Tank capacity (maker's figure) . . . 10 gal.

Brakes

Pedal pressure, deceleration and equivalent stopping distance from 30 m.p.h.

lb.	g	ft.
25	0.33	91
50	0.68	44
70	0.95	31½
Handbrake	0.43	70

Fade test

20 stops at ½g deceleration at 1 min. intervals from a speed midway between 30 m.p.h. and maximum speed (=63.1 m.p.h.)

	lb.
Pedal force at beginning	38
Pedal force at 10th stop	37
Pedal force at 20th stop	39

Steering

Turning circle between kerbs:

	ft.
Left	27⅓
Right	28⅓

Turns of steering wheel from lock to lock . . 3.9
Steering wheel deflection for 50 ft. diameter circle . . . 1.05 turns

Clutch

Free pedal movement . . . = ½ in.
Additional movement to disengage clutch completely . . . = 3 in.
Maximum pedal load . . . =22 lb.

Speedometer

Indicated	10	20	30	40	50	60	70	80	90
True	9	20	30	39½	49½	58½	69	79	89

Distance recorder . . . 1½% fast

Weight

Kerb weight (unladen with fuel for approximately 50 miles) . . . 18.2 cwt.
Front/rear distribution . . . 55/45
Weight laden as tested . . . 21.9 cwt.

Parkability

Gap needed to clear a 6ft. wide obstruction parked in front:

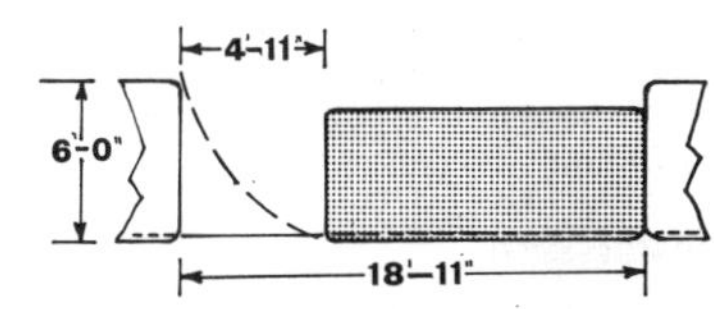

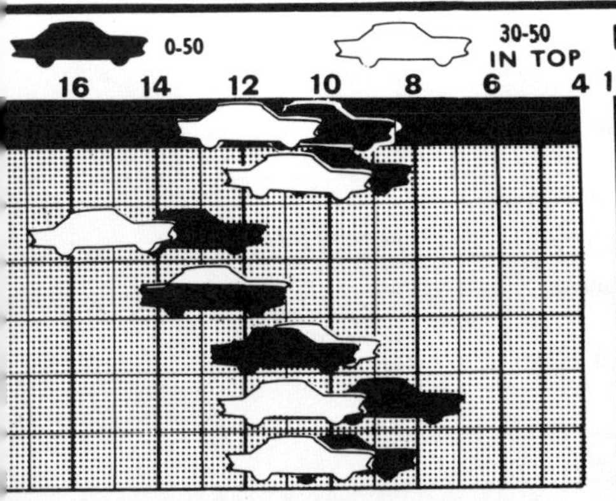

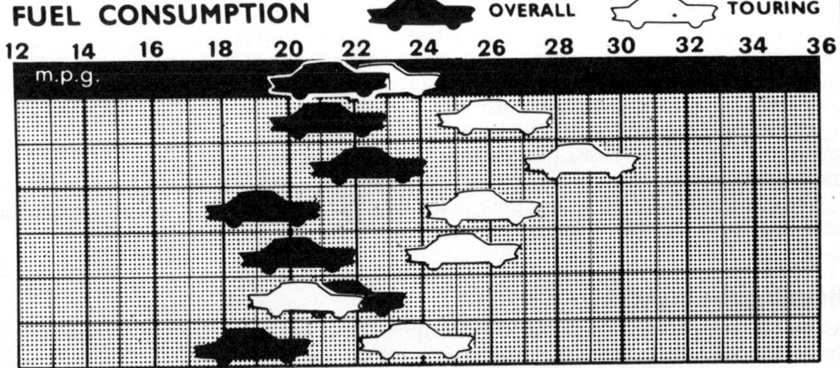

Wooden cappings and facia make the 1600E look quite luxurious inside; reclining seats are standard—they hold well on corners too. Central locker is of useful size.

Ford Cortina 1600E *continued*

speed). This maximum is usefully higher than the 91.5 m.p.h. of our 1963 Cortina GT and very good for a 1.6-litre saloon which, incidentally, has a higher drag factor than the previous Cortina shape. However, the extra power has only just got the better of greater weight and a higher first gear, as witness the 0—60 time of 11.8 s. compared with 12.1 s. for the original 1500GT. These figures certainly confirm the impression of "instant acceleration" you feel when surging past slow traffic in the high second gear.

If you take the 4,000 r.p.m. harshness seriously (and it isn't really that obtrusive) you will probably do most of your motoring on the primary choke of the compound Weber, to the benefit of fuel consumption; although the transition between the chokes is not so easy to feel through the accelerator as it used to be, such motoring should return 25–28 m.p.g. while more spirited driving will drop the figure nearer to our overall 23.1 m.p.g. On a 10-gallon tank this allows only about 200 miles per tankful of the recommended four-star fuel; the engine seemed quite happy on the lower (97 octane) end of this scale.

Transmission

The gearbox is really excellent with unbeatable synchromesh on all four ratios; however fast you push the solid lever through the gate the movement is always smooth and the next gear seems automatically to pick up at exactly the right revs. The ratios aren't particularly close (as in the original Lotus Cortina) but they suit almost any conditions on the road with particularly useful second gear performance up to 50 m.p.h. First gear is good for over 30 m.p.h. but not for a start on a 1-in-3 hill as the light clutch, which juddered slightly when inching forward either when parking or in traffic, just slipped.

There is a little gear whine noticeable in the three indirects but once into top there is no whine at all from either the gearbox or back axle.

Handling and brakes

As well as being lower than on the GT, the front suspension on the 1600E has a zero castor angle which makes the steering feel a little dead, even sticky, in the straight ahead position; once into a corner though, the self aligning torque of the big India Autobands gives plenty of useful feel and on wet roads you know exactly how much adhesion is left.

With a smaller leather rimmed wheel the steering feels pleasantly direct, just over one turn being needed for a 50 ft. circle; but it also feels a little heavy on tight turns beyond the crossed-arms stage.

The grip on dry roads is very good and the car feels extremely stable when cornering at high speeds; it is even more impressive on slippery surfaces, particularly those wet and greasy ones you find in the farming countryside at this time of year. It will take corners at speeds that you think must provoke a tail slide yet you

Safety check list

Steering assembly	
Steering box position	On bulkhead
Steering column collapsible	No
Steering wheel boss padded	No
Steering wheel dished	Yes
Instrument panel	
Projecting switches	Yes
Sharp cowls	No
Padding	Top and bottom of facia and parcel shelf
Windscreen and visibility	
Screen type	Zone toughened
Pillars padded	Covered
Standard driving mirrors	Interior
Interior mirror framed	Yes
Interior mirror collapsible	Yes
Sun visors	Soft, collapsible
Seats and harness	
Attachment to floor	Runners bolted to floor
Do they tip forward?	No
Head rest attachment points	No
Back of front seats	Padded frame
Safety harness	Lap and diagonal
Harness anchors at back	No
Doors	
Projecting handles	Yes
Anti-burst latches	Yes
Child-proof locks	Yes—on rear doors

come out thinking: "Gosh, I could have gone through that even faster". On public roads, it is usually the driver's nerve, not the tyres' adhesion that sets the limit which makes it all very safe. On wet roads you can produce a final gentle oversteer by using full power in third but you have to try hard in second on dry roads to get the tail going at all, and even then the slight side-step is more or less self-correcting. The behaviour is much the same in the wet as the tail never swings wildly but just floats gently aside and can be held very easily. This is the first test car we have had on India Autobands and they deserve a lot of credit for the good wet roadholding of the 1600E; only when trying very hard at MIRA did they squeal. Radius arms help, too, as there was never any tendency to tramp, even during our standing starts. These used to provoke the most ear-shattering thumps from the back axle on previous GTs but good suspension design has cured this and also improved the handling, while efficient damping keeps the car on an even keel in S bends and there is little roll even on long fast corners. When trying hard and enjoying this sort of fast motoring you tend to forget deficiencies in riding comfort.

The brakes feel pleasantly firm and don't have the oversensitivity of some servo-assisted systems which makes the well placed pedal an ideal leaning post during heel-and-toeing. In our fade test there was no deterioration at all but the watersplash considerably reduced braking efficiency and several stops were needed before this was restored. The rear wheels locked first during our maximum stop and slewed the tail a bit but 0.96g was good on the surface; 0.43g was achieved by the handbrake in a locked rear wheel stop and it could hold the car easily on a 1-in-3 hill.

Continued on the next page

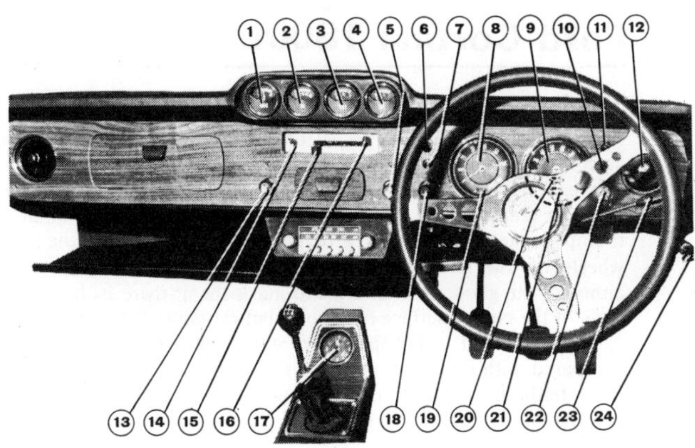

1, ammeter. 2, oil pressure gauge. 3, temperature gauge. 4, fuel gauge. 5, cigar lighter. 6, panel light. 7, LH indicator tell-tale. 8, rev. counter. 9, speedometer. 10, RH indicator tell-tale. 11, light switch. 12, facia vent control. 13, choke. 14, two speed fan. 15, heater direction control. 16, heater temperature. 17, clock. 18, wiper/washer. 19, generator tell-tale. 20, total mileage recorder. 21, main beam tell-tale. 22, ignition/starter key. 23, indicator/flasher/dipper stalk. 24, bonnet release.

Specification

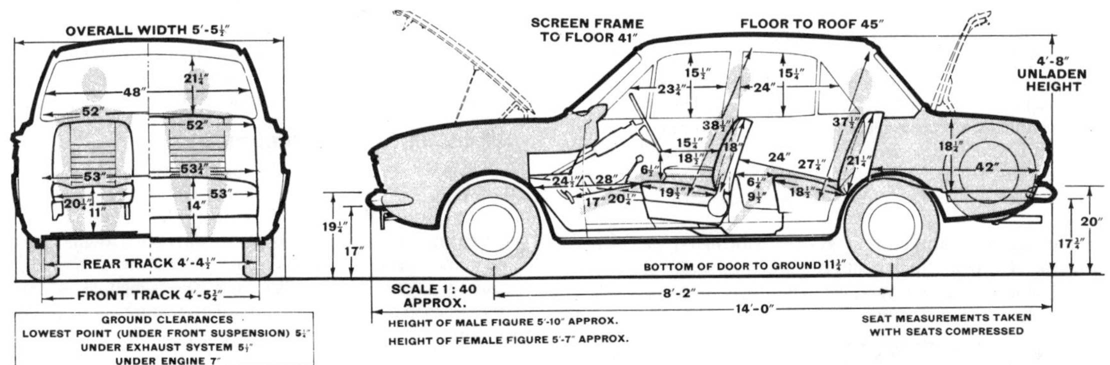

Engine

Cylinders	4
Bore and stroke	81.0 mm. x 77.6 mm.
Cubic capacity	1,599 c.c.
Valves	Pushrod o.h.v.
Compression ratio	9.0:1
Carburetter	Weber 32DFM mechanical compound
Fuel pump	AC mechanical
Oil filter	Fram or Tecalemit full flow
Max. power (net)	88 b.h.p. at 5,400 r.p.m.
Max. torque (net)	96 lb.ft. at 3,600 r.p.m.

Transmission

Clutch	Borg and Beck s.d.p. 7.54 in. dia.
Top gear (s/m)	1.00
3rd gear (s/m)	1.40
2nd gear (s/m)	2.01
1st gear (s/m)	2.97
Reverse	3.32
Final drive	Hypoid bevel 3.91:1
M.p.h. at 1,000 r.p.m. in:—	
Top gear	17.2
3rd gear	12.3
2nd gear	8.6
1st gear	5.8

Chassis

Construction	Unitary

Brakes

Type	Girling disc/drum
Dimensions	Discs 9⅝ in. dia. Drums 9 in. dia.
Friction areas:	
Front	20.64 sq.in. of lining operating on 189.5 sq.in. of disc
Rear	48.0 sq.in. of lining operating on 98.8 sq.in. of drum

Suspension and steering

Front	Independent; MacPherson strut with lower wishbone incorporating anti-roll bar; coil springs
Rear	Live axle with leaf springs and upper radius arms
Shock absorbers:	
Front	Telescopic with strut
Rear	Telescopic
Steering gear	Burman recirculating ball
Tyres	India Autoband 165-13
Rim size	5½ J-13

Coachwork and equipment

Starting handle	No
Jack	Screw pillar
Jacking points	Two each side under door sills
Battery	12 volt negative earth, 38 amp hrs. capacity
Number of electrical fuses	None
Indicators	Self-cancelling flashers
Screen wipers	Single speed, self parking
Screen washers	Manual plunger
Sunvisors	Two
Locks:	
With ignition key	Driver's door and boot
Interior heater	Fresh air with cold air facia vents
Extras	Radio
Upholstery	Pvc
Floor covering	Carpet
Alternative body styles	None

Maintenance

Sump	6.2 pints SAE 10W/30
Gearbox	2.1 pints SAE 80 EP
Rear axle	2 pints SAE 90 EP
Steering gear	EP 90
Cooling system	11.4 pints (2 drain taps)
Chassis lubrication	None
Minimum service interval	6,000 miles
Ignition timing	8° b.t.d.c.
Contact breaker gap	0.025 in.
Sparking plug gap	0.023/0.027 in.
Sparking plug type	Autolite AG22A
Tappet clearances (hot)	Inlet 0.012 in.; Exhaust 0.022 in.
Valve timing:	
Inlet opens	27° b.t.d.c.
Inlet closes	65° a.b.d.c.
Exhaust opens	65° b.b.d.c.
Exhaust closes	27° a.t.d.c.
Front wheel toe-in	0.12–0.18 in.
Camber angle	1°
Castor angle	0°
King pin inclination	7° 56′
Tyre pressures:	
Front	24 p.s.i.
Rear	24 p.s.i.

Ford Cortina 1600E *continued*

Comfort and controls

Although the Cortina Lotus suspension is fine for earholing in safety you have to accept sacrifices in the ride with what is a fairly conventional suspension design. At town speeds the radial tyes transmit almost every bump as a remote but nevertheless audible thump, bigger disturbances generate other thumps from the rear wheels via the rather rigid radius arms. You feel all these too, although the good seats do a lot to mask them; there is, however, none of the coarse surface road roar that cross ply tyres create.

Out of town at higher speeds the stiffer springing on the 1600E, compared with the 1600, makes the car follow the road contours fairly faithfully giving a restless, if not uncomfortable, ride on all but the smoothest surfaces, though good damping prevents any float.

We all liked the seating position; it is comfortable for all sizes, gives a commanding view over the bonnet and there are no blind spots unless the fixed quarter light gets dirty. The cushion is rather horizontal and therefore lacking in thigh support, but the back rest is well shaped for side location and adjusts through a wide range of angles. The main criticism is that the low steering wheel is too close to the knees of crossed arm twirlers or those who heel and toe.

Once you have adjusted the lap and diagonal seat belt so that the buckle rests on your hip, rather than behind the seat, it is quite easy to fasten; the handbrake gets a little far away when the belt is tight but all other controls are within easy reach; inertia reels are available for those who follow driving test procedure and grab the handbrake at every stop. The back seat (for which there are no belt anchorages) will take three adults fairly comfortably and two very easily with plenty of knee and shoulder room, even with the front seats fully back.

We have already mentioned the engine and wind noise; the latter stems from the door seals and starts as a steady rush from about 40 m.p.h., building up a little as speed increases. If you adjust the radio, a good Ford extra, for comfortable listening at 40 m.p.h. you can't understand speech at 70 m.p.h.; many cars are worse but an E should be better. Improved door seals and a little engine smoothing would make this an effortless 80—90 m.p.h. autobahn cruiser; as it is it may well get tiring with prolonged over-70 use.

Retaining the original Aeroflow system with better facia vent controls, the Cortina equipment is more versatile than the simplified Escort or Zephyr versions. The direction lever gives a degree of volume control if you keep it between "screen" and "off", although the handbook makes no mention of this fact; most people will just divert the greater flow to the screen if their legs are getting too warm and use the fresh air vents as coolers. It is an extremely good system which needs little readjustment with changing road speed and it keeps the windows well demisted. A two-speed booster fan is also fitted.

On the 1600E a pair of Wipac spot lamps supplement the existing lights on main beam to give a simple "four-eyed" conversion; the result is that main beam is so bright and strong that dip is almost yellow by comparison giving the effect of plunging you into relative darkness. The two-speed wipers don't lift off at speed, clean the screen well and sweep right up to the edge of the right hand pillar.

1, battery. 2, brake fluid reservoir. 3, starter solenoid. 4, oil filler cap. 5, radiator cap. 6, dipstick. 7, junction box. 8, washer reservoir.

MAKE: Ford. MODEL: 1600E. MAKERS: Ford Motor Company Ltd., Dagenham, Essex.

Fittings and furniture

The interior design of the 1600E is identical to that of the GT except for wood cappings all round the waistline and on the facia where the extra thickness is used to recess the two lighting switches, panel and side/head. The other controls are well placed within reach including the combined flasher/dipper/indicator stalk, and the auxiliary instruments mounted on top are easy to glance at. A clock is fitted on the forward end of the gear lever console.

While the car is itself safe to drive by virtue of its swervability, acceleration and brakes, some of the facia and door protrusions look a bit sharp and unlikely to collapse if hit by flying bodies in the second collision.

A further criticism of the door design is that you should be able to unlock your passenger's door first, an elementary courtesy which could perhaps be forgiven if the owner had a cheaper model than the 1600E.

Upholstery in pvc, and carpets, make the 1600E easy to keep clean but the light-coloured seats in our car may well get dirty rather quickly if E stands for Enthusiast. A glove locker is provided in all the Cortinas but only the GT, 1600E and Lotus versions get a useful central locker cum arm rest. Most of the luggage can go in the vast boot which takes 10.9 cu. ft. of our test luggage, possibly a shade less than a GT which doesn't have the $5\frac{1}{2}$J spare wheel (now hidden under a cover).

Accessibility and maintenance

The 1600E has an interior bonnet release, a fact which confused more than one mechanic checking the oil—it's usually the button in the centre of the grille. Once open, the bonnet is propped on a convenient strut and there is plenty of room around the engine for the home mechanic to do most of his own servicing. This is required every 6,000 miles and the schedule is particularly comprehensive, the only additional features which don't need attention every 6,000 miles being a new air cleaner element and the checking of front wheel toe-in. So it is no longer a question of the home mechanic doing the small services between the major ones—they are all major but less frequent. The standard toolkit will be no help at all for this as you still only get wheel changing equipment; this can now be clipped to the floor instead of sliding around. **M**

Insurance

AOA Group rating	4
Lloyd's	4

Maintenance summary

Every 6,000 miles change engine oil and filter, clean air cleaner, check and adjust valve clearances, clean crankcase emission valve, clean oil filler cap, clean fuel filter bowl, lubricate distributor and generator rear bearing, check and adjust distributor points, clean distributor cap and coil, clean and reset plugs, adjust fan belt and tighten generator bolts, check battery level, check radiator level, check washer operation, top-up gearbox and rear axle, check rear spring U-bolts and inserts, check front suspension cross-member bolts and front suspension/steering joint gaiters, check and top-up clutch and brake reservoirs, check pads and brake shoes for wear, inspect brake hoses, top up steering box, check and adjust front wheel bearings, lubricate hand brake cable, door locks, bonnet safety catch and all oil can points, check operation of all controls instruments and lights, check seat belts for security and wear, road test car and adjust carburetter and ignition if necessary.

Every 18,000 miles renew air cleaner, check front wheel toe-in, repack and adjust front wheel bearings.

Every 36,000 miles renew all brake seals hydraulic fluid and flexible brake hoses.

CAR and DRIVER ROAD TEST

Ford Cortina 1600 GT

It's surely one of the superest things English Ford has ever hatched—but a "micro-Super Car"?

PHOTOGRAPHY: GENE BUTERA

Is there such a thing as a Super Car from a disadvantaged home? Is the Cortina GT a Super Car? Is England disadvantaged? Normally when we think of super cars, we think of fat-tired, intermediate-sized American sedans with enough torque to yank the Washington Monument out by the roots. Ford of England disagrees and from every orifice of its works in Dagenham has come spewing a hopped-up little saloon which it insists is a micro-Super Car and surely one of the superest things English Ford has ever hatched.

Perhaps we should establish a Super Saloon category.

The car is based on the Cortina 1600 and is known to the world quite simply and totally without guile as the Cortina 1600 GT. Cortinas have been salamander-like in recent years; they've been the Model C to Model A and T lovers, Lotus Cortinas to followers of USRRC and Trans-Am racing—put a Cortina down somewhere and you will see it take on protective coloration. All right, what's the GT all about? Our first meeting with the test car was a howling disappointment. It looked like anything but a super GT car. It just sat over in the corner of the garage wearing a used tomato juice red paint job—Dragoon Red says the Cortina color chart but it must have been a dragoon with a very serious blood problem. Anyway, the thing was perched on a set of Goodyear 165 SR 13 radial ply tires, the kind that have a tread pattern like motorcycle scrambler tires, not looking as though it had even the ambition to move. GT? Well, it did have a GT *sign* embroidered on one buttock so, most likely, we weren't involved in a case of mistaken identity.

Our impression changed the very moment we set eyes on the instrument panel. It's magnificent—better than most real sports cars. Right in front of the driver is a large, round speedometer and tachometer. Over toward the center of the panel are four more round gauges, fuel, oil, temp, and amps—just like an instrument panel should be. Each gauge has white numbers on a black background. No trick stuff. Just straight-forward information sources.

The normal sequence of our road test procedure indicated some kind of real involvement about this time. Like get in the car and point it toward the horizon—great idea, but we had our shoes on. What's that got to do with anything? The Cortina has the same number of foot pedals as any other manual transmission car—clutch, brake and accelerator; moreover it's a roomy little sedan, so there's plenty of space for pedals. Why is it, then, that the brake pedal is so close to the transmission hump that a size 10½D wing-tip can't slip past? Super Cars, even micro-Super Cars don't make the program if you can't get your foot on the gas pedal. All this begins to say what's bad—and what's good—about the Cortina.

There are times we're convinced that British automotive designers have less imagination than an anvil, and the Cortina is a case in point. A decidedly British flavor surrounds the Cortina—even the heater speaks with a British accent, the same pulsating whir we've heard from British blowers for the past dozen years; and the drivetrain broadcasts the same gravelly whine that characterized English sports cars after World War II. Ford might just as well have painted the Union Jack on its bonnet, because the Cortina's origin couldn't be more obvious. This is not to say a foreign car has to be Swedish or German or Italian or (lately) Japanese to be worthwhile—but 12 years? Can't we expect some awareness of even little problems like heater noise after 12 years? Please?

The big news about the Cortina and the deciding factor in arousing our interest enough to do a road test was the car's redesigned engine. From the name, Cortina 1600 GT, you may have suspected the displacement is now 1600cc, up 100cc from past Cortinas. Your suspicion is well founded. What Ford is really crowing about, though, is the new crossflow cylinder head for the in-line Four. It's still of the pushrod operated, overhead valve type but now the intake ports are on one side of the head and the exhaust ports on the other. "No corners for the gases to turn—they go straight in and out," says Ford, celebrating the whole philosophy of crossflow and the wonders it does for breathing. Of at least equal interest are the combustion chambers, now dish-shaped cavities in the tops of the pistons with a flat surface on the head, just the reverse of the conventional system. A similar idea was used on the 348-409 series Chev-

The Cortina GT feels tough, like a midget super stocker, so we treated it accordingly. It didn't even flinch—it just kept coming back for more.

rolet V-8 starting in 1958 and is frequently seen in Diesels.

Like most foreign manufacturers that sell cars in the United States, Ford controls exhaust emission in the Cortina with an air pump. Blowing fresh air into the exhaust port causes the unburned hydrocarbons and carbon monoxide to burn in the exhaust manifold rather than be exhaled out the tailpipe like bad breath. This process results in abnormally high exhaust manifold temperatures which may shorten the life of the exhaust header tubing.

All this has been carping to some degree, because it's annoying to see someone set out to build a *car* rather than a module and lapse into diverse idiocies. But the GT *is* a car and there are certain microsplendors about it as a result. While the basic engine is shared between the standard Cortina and the Cortina GT, the GT version gets some help in the form of a twin-choke Weber carburetor, higher compression ratio, and an impressive looking tri-style exhaust header of steel tubing. Ford advertises the output of its high-potency GT engine as 89 hp at 5500 rpm, and we believe it. With less than 200 miles on the odometer when we made our timed runs, the GT would do a standing quarter mile in 18.6 seconds at 73.2 mph. The scrambler tires weren't much for bite on the asphalt which made the ET a tenth or two worse than it could have been. It's surprising to note how close the factory performance figures correspond to those of the test car. Ford claims 0–60 mph in 12 seconds and a top speed of 95 mph. The test car did 0–60 mph in 12.2 seconds and was still accelerating at 92 mph when we shut off to avoid driving off the end of our test strip. Given more room it would have made 95 mph and when fully broken in 97 mph is quite likely—maybe even 100. Our confidence in the words of auto manufacturers is somewhat restored now—they don't all speak with forked tongues after all.

As vile as the Formula Vee racers tell us drag racing is, we have to admit we enjoyed the drag strip part of the Cortina test. The car feels *tough*, like a midget super stocker, so we treated it accordingly. What a gas. Rev it up. Pop the clutch. The rear tires buzz a little bit, just like the real drag racers, and the Cortina is off. Then it's time to play Dick Landy with the shift lever. OK. Accelerator to the wood, stab the clutch and pull the lever to the next gear—all in a micro-second; true communion between man and machine. Again and again for third and fourth, each time at 6000 rpm. The Cortina doesn't even flinch—just comes back for more. Its shifting mechanism is very precise. Never any confusion about what the hidden portion of the shift lever is doing or the feeling that it might even come off in your hand. There aren't many imported sedans in which we would even try this rude behavior because it's embarrassing to bring back a test car in a basket, but the Cortina just invites floggers like us to do our damndest. Even though the transmission is super noisy its virtues outweigh its faults by a whole lot.

The Cortina contains no outstanding surprises in the suspension department. A garden variety MacPherson strut is used in front with the assistance of an anti-sway bar. Semi-elliptic leaf springs hold up the rear. Unhappily, the wheels are only four inches wide. That's the going width in the econo car business, but the GT is supposed to be a notch above plebian transportation and wider wheels (as well as hop-up equipment) *are* available for the car. English Ford doesn't maintain a full time competitions director in the U.S. any more, but specialty firms like Northern California's WinkSpeed offer bolt-ons to make the little car into a genuine pocket terror.

It's a good thing. When it comes to handling, the Cortina needs help. It's predictable enough to suit even long-range planners, but its cornering speeds are not what we expect of a car with sporting flavor. The inside rear wheel unloads, even in large radius turns, allowing that wheel to spin uselessly. How can you get around a corner in a hurry when you can't put the power to the ground? All the while it's trying to hang its tail out but it takes a real effort to get it going fast enough to do the job.

The Cortina isn't too sure-footed during quick lane changes either. Although not dangerous, it tends to wag its tail like a station wagon. Ford could well spend more development time on the Cortina's suspension—and more roll stiffness in the front would be a good place to start.

The Cortina is bigger than you would expect. Not a whole lot, mind you, but enough to be noticeable. After checking out the dimensions we found the Cortina's 98-inch wheelbase is about three inches longer than its Opel, Datsun and Toyota competitors and that its 168-inch overall length is 3.3 inches longer than an Opel and almost six inches longer than a Datsun or Toyota. The Cortina has plenty of room for four even though it could never be called spacious by American standards.

Room or no room, the same guy who's been darting around the British motor industry sneaking in dividers in interior design whenever possible has struck again. In the Cortina his pack-rat complex has gone too far. He's got storage compartments in his storage compartments—or so it seems. The GT has a conventional glovebox in the dash. Everybody has those, right? That act is followed by a parcel tray under the dash on the right side. Not really too surprising because parcel trays have become a European tradition. Now, since the Cortina GT is a sporty-type car, and sporty cars are recognized by chrome proclamations on the outside and consoles between the front bucket seats on the inside, the Cortina had to have a console, right? Motivated by a desire to build a truly good sporty sedan, the Ford people did a fantastic job on the console. Obviously any console worth its space has a storage compartment, so the Cortina has a total of *three* bins for its occupants to fill. But the bin in the console is a masterpiece. It has a padded lid which serves as a center armrest for the front seat passengers, the opening is surrounded with bright metal trim and there is even a friction apparatus which holds the lid open in an infinite number of positions. To give a console an official appearance it probably should contain at least one instrument—and the more important the instrument the better. You're not going to catch Ford napping there, either. Ford followed the prescription by putting a clock in the Cortina GT's console—and next year, who knows, it may be joined by an anemometer.

In the unlikely instance that you've been dazzled by the clock, the tachometer brings you back to reality—British reality. Right there on the face of the tach in white letters appear the words *negative earth.* One way or another, there will always be an England.

When it comes to schemes and devices to actuate the horn, the Cortina ranks with the best of them. Would you believe the turn signal lever telescopes and just as the lever reaches its shortest length the horn goes off? The end of the turn signal lever makes a middling small target when you're frantically trying to honk the horn but we've seen worse systems.

Ford has chosen a shotgun approach to interior ventilation—flow-through system and vent windows too. That's taking no chances, like wearing suspenders and a belt. It works, though, and it works well and that's where the points get counted.

Now that the safety regulations are in effect

Text continued on page 10
Specifications overleaf

LOCATION: KNOTTS BERRY FARM, BUENA PARK, CALIFORNIA

GRIST MILL

FORD CORTINA 1600 GT

Importer: English Ford Line Operation
3000 Schaefer Road
Dearborn, Michigan

Number of dealers in U.S.: 830

Vehicle type: Front-engine, rear-wheel-drive, 4-passenger

Price as tested: $2337.17

(Manufacturer's suggested retail price, including all options listed below, Federal excise tax, dealer preparation and delivery charges; does not include state and local taxes, license or freight charges)

Options on test car:

Interior decor group ($17.42), exterior decor group ($13.62), radial ply tires ($43.19)

ENGINE

Type: water-cooled 4-in-line, cast iron block and head, 5 main bearings
Bore x stroke 3.189 x 3.056 in, 81.0 x 77.5 mm
Displacement ... 97.6 cu in, 1600cc
Compression ratio ... 9.6 to one
Carburetion ... 1 x 2-bbl Weber
Valve gear ... Pushrod-operated overhead valves, mechanical lifters
Power (SAE) ... 89 bhp @ 5500 rpm
Torque (SAE) ... 102.5 lbs/ft @ 4000 rpm
Specific power output ... 0.91 bhp/cu in, 55.6 bhp/liter
Max. recommended engine speed ... 6000 rpm

DRIVE TRAIN

Transmission ... 4-speed manual, all-synchro
Clutch diameter ... 7.54 in
Final drive ratio ... 3.90 to one

Gear	Ratio	Mph/1000 rpm	Max. test speed
I	2.97	5.8	35 mph (6000 rpm)
II	2.01	8.6	52 mph (6000 rpm)
III	1.40	12.3	74 mph (6000 rpm)
IV	1.00	17.3	93 mph (5400 rpm)

DIMENSIONS AND CAPACITIES

Wheelbase ... 98.0 in
Track ... F: 52.5 in, R: 51.0 in
Length ... 168.0 in
Width ... 64.9 in
Height ... 55.0 in
Ground clearance ... 5.2 in
Curb weight ... 2020 lbs
Test weight ... 2170 lbs
Weight distribution, F/R ... 54.3/45.7%
Lbs/bhp (test weight) ... 24.4
Battery capacity ... 12 volts, 58 amp/hr
Generator capacity ... 264 watts
Fuel capacity ... 12.0 gal
Oil capacity ... 3.1 qts
Water capacity ... 6.7 qts

SUSPENSION

F: Ind., MacPherson strut, coil springs, anti-sway bar
R: Rigid axle, semi-elliptic leaf springs

STEERING

Type ... Recirculating ball
Turns lock-to-lock ... 4.5
Turning circle ... 30.8 ft

BRAKES

F: ... 9.63-in. solid discs
R: ... 9.0 x 1.75 cast iron drums
Swept area ... 285.6 sq in

WHEELS AND TIRES

Wheel size and type ... 13 x 4.0-in, stamped steel, 4 bolt
Tire make, size and type ... Goodyear 165 SR 13 radial ply, tubeless
Test inflation pressures ... F: 24 psi, R: 24 psi
Tire load rating ... 760 lbs per tire @ 24 psi

PERFORMANCE

Zero to	Seconds
30 mph	3.3
40 mph	5.6
50 mph	8.3
60 mph	12.2
70 mph	16.7
80 mph	23.3

Standing ¼-mile ... 18.6 sec @ 73.2 mph
80–0 mph panic stop ... 264 ft (.81 G)
Fuel mileage ... 19–23 mpg on premium fuel
Cruising range ... 228–276 mi

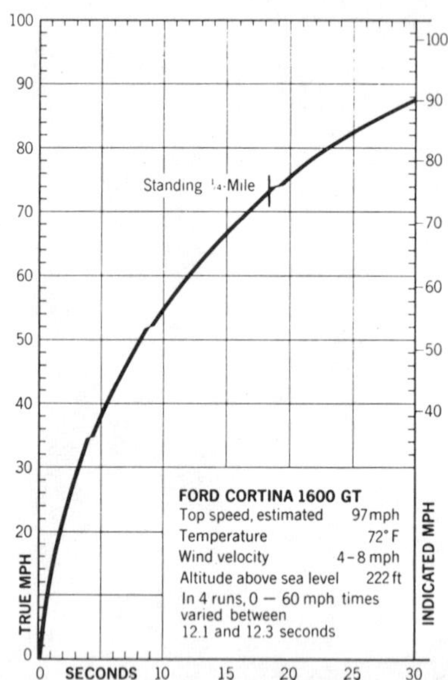

CHECK LIST

ENGINE

Starting ... Fair
Response ... Fair
Vibration ... Fair
Noise ... Fair

DRIVE TRAIN

Shift linkage ... Very Good
Synchro action ... Very Good
Clutch smoothness ... Good
Drive train noise ... Poor

STEERING

Effort ... Good
Response ... Very Good
Road feel ... Very Good
Kickback ... Very Good

SUSPENSION

Ride comfort ... Fair
Roll resistance ... Good
Brake dive ... Good
Harshness control ... Fair

HANDLING

Directional control ... Good
Predictability ... Very Good
Evasive maneuverability ... Very Good
Resistance to sidewinds ... Fair

BRAKES

Pedal pressure ... Good
Response ... Fair
Fade resistance ... Very Good
Directional stability ... Good

CONTROLS

Wheel position ... Fair
Pedal position ... Poor
Gearshift position ... Very Good
Relationship ... Good
Small controls ... Very Good

INTERIOR

Ease of entry/exit ... Very Good
Noise level (cruising) ... Fair
Front seating comfort ... Good
Front leg room ... Good
Front head room ... Good
Front hip/shoulder room ... Good
Rear seating comfort ... Fair
Rear leg room ... Fair
Rear head room ... Good
Rear hip/shoulder room ... Very Good
Instrument comprehensiveness ... Excellent
Instrument legibility ... Excellent

VISION

Forward ... Very Good
Front quarter ... Good
Side ... Good
Rear quarter ... Good
Rear ... Excellent

WEATHER PROTECTION

Heater/defroster ... Good
Ventilation ... Good
Weather sealing ... Very Good

CONSTRUCTION QUALITY

Sheet metal ... Good
Paint ... Good
Chrome ... Fair
Upholstery ... Good
Padding ... Very Good
Hardware ... Fair

GENERAL

Parking and signal lights ... Very Good
Wiper effectiveness ... Fair
Service accessibility ... Fair
Trunk space ... Excellent
Interior storage space ... Excellent
Bumper protection ... Good

CONTINUED FROM PAGE 22
the Ford and certainly makes it look different. Inside, there is lots of polished wood everywhere which some people might consider old-fashioned, but it attracted much admiration, both from passengers and casual passers-by. The driver has a splendid display of proper round instruments and a leather-covered steering wheel to make him feel good. Naturally, the power unit is the GT version of the new 1600.

On the road, the 1600E is noticeably much livelier than previous Cortinas. These cars have sometimes been criticized for lacking "punch" in the middle ranges, but nobody can ever say that again, and it is difficult to believe that the engine has a capacity of less than 2 litres. The car is certainly ideal for this country with its 70 mph limit, and third gear, which will just exceed this speed, gives really vivid acceleration to regain the legal cruising speed. The new higher second gear is an absolute delight—one could hardly imagine that a single alteration would do so much for a car—and it is no longer necessary to remain in third round sharp corners.

The car will cruise with great ease at 70 to 80 mph, which is its best speed. From 85 mph upwards, the engine is obviously doing more work and it is considerably noisier, but it will eventually attain about 95 mph, which is indicated as well over 100 mph on the speedometer. Incidentally, an executive would surely need a trip reading on his speedometer for timing his individual journeys. Driven hard, the 1600E records not less than 25 mpg in open country.

In town, the performance is not quite so outstanding. This is partly due to the clutch, which does not match the excellence of the gearbox, tending to be heavy in action with rather a long travel on the pedal. The engine is not very flexible at low speeds, and it seems best to forget about top gear in London. Above 30 mph, the car suddenly comes alive, and it retains a useful performance even when one is too lazy to use the gearlever. Certainly, most owners will prefer to employ the gearbox to the full, for the remote control gearlever is a joy to handle.

The roadholding, assisted by radial ply tyres, is superb—no lesser word suffices. On wet roads or dry, the driver gains complete confidence, and the handling is absolutely predictable. The steering is quick and precise, feeling so right all the time, and the wheels seem glued to the road, irrespective of surface. The ride is not soft, and there is some up and down movement on certain country roads, but under more normal conditions the driver and his passengers travel in comfort. The brakes are progressive and stand up well to quite hard driving without smelling excessively hot.

The Ford heating and ventilation system has often been praised. It is certainly excellent, and this is one of the few cars which can be warm inside without being stuffy. It actually seems possible to control the heat, without the "all or nothing" effect which most heaters give. It also adds to the pleasure of riding in the car to be seated in such a well-furnished interior.

It was Laurence Pomeroy the elder, I think, who said that if you have to use a stopwatch to find out whether you have made an improvement or not, you have not made a worthwhile improvement. In their crossflow engine, Fords have an improvement that needs no stopwatch for proof. Particularly under present day road conditions, the Cortina, in all its versions, is now a far better car. When this new performance is allied with superior comfort and appearance, the Cortina becomes a most desirable possession and I predict an enthusiastic demand for the 1600E, the Cortina for the VIP.

SPECIFICATION AND PERFORMANCE DATA

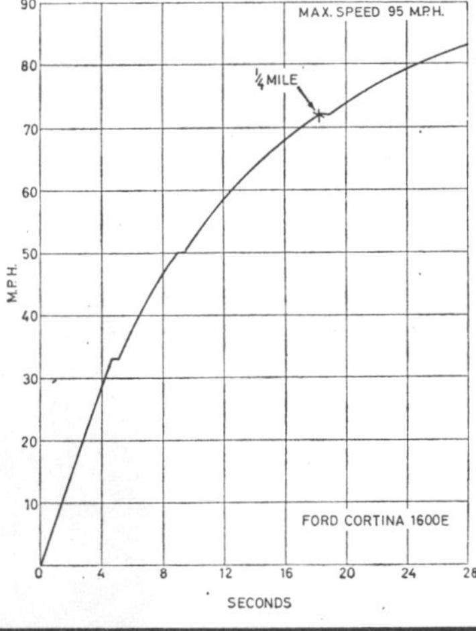

Car tested: Ford Cortina 1600E 4-door saloon, price £1020 18s 11d including PT.

Engine: Four-cylinders, 80.98 mm x 77.62 mm (1599 cc). Pushrod operated overhead valves. Compression ratio 9.2:1. 93 (gross) bhp at 5400 rpm. Weber twin-choke down-draught carburetter. Lucas coil and distributor.

Transmission: Single dry plate clutch. 4-speed all-synchromesh gearbox with central remote control, ratios 1.0, 1.40, 2.01, and 2.97:1. Open propeller shaft. Hypoid rear axle, ratio 3.90:1.

Chassis: Combined pressed-steel body and chassis. Independent front suspension by MacPherson struts with lower wishbones and anti-roll bar. Burman recirculating ball steering gear. Live rear axle on semi-elliptic springs and trailing links. Telescopic dampers all round. Bolt-on pressed steel wheels with chromium decoration, fitted 165 x 13 radial ply tyres. Girling disc front and drum rear brakes.

Equipment: 12-volt lighting and starting. Speedometer. Rev counter. Ammeter. Oil pressure, water temperature, and fuel gauges. Clock. Heating, demisting and ventilation system. Windscreen wipers and washers. Flashing direction indicators. Cigar lighter. Paired driving lamps. Reversing light. Radio (extra).

Dimensions: Wheelbase, 8 ft 2 ins; track (front), 4 ft 4.5 ins; (rear), 4 ft 3 ins; overall length, 14 ft; width, 5 ft 4.9 ins; weight, 18 cwt 16 lbs.

Performance: Maximum speed, 95 mph. Speeds in gears: third, 72 mph; second, 50 mph; first, 33 mph. Standing quarter-mile, 18.3 s. Acceleration: 0-30 mph, 4.2 s; 0-50 mph, 9 s; 0-60 mph, 12.4 s; 0-80 mph, 25 s.

Fuel consumption: 24 to 28 mpg.

As we always said, a new head would make all the difference...

CROSS FLOW CORTINA

One of our favorite cars now has that snap, crackle and pop, that top-of-the-morning feeling, after languishing in early middle age with slight overweight problems. Gentlemen, meet the new Cortina 1600 GT.

FULL ROAD TEST OF OUR

LET the buyer beware! There is only one real way to pick the 1600 cc cross-flow Cortina GT — and that's by engine number. There are no badges or tell-tales anywhere — except for two things: a forward extension on the transmission tunnel console that turns up at its end and carries large GT letters and a badge at the rear. But a smart used car salesman could easily fit two of those to a 1500 GT, and nobody would be the wiser.

Why all the fuss about identification? Because nobody in his right mind would take a 1500 GT after sampling the larger, cross-flow, bowl-in-piston engine. When the new look Cortinas arrived last August, the GT was found to be suffering from slight debility brought on by increased girth. Despite Ford's protestations the extra weight took the edge off what was really quite an adequate performance for the money. But no stock 1500 GT could cut a quarter in much less than 19.5 secs, nor get much over 93 mph, so the GT badge started to look a little tarnished. And then again, where in 1964-65 the GT was just about the

COVER CAR THAT HAS THE HONOR OF CARRYING OUR 15th ANNIVERSARY

best performance package for the money, it has since been overtaken by a number of other cars, including S-type Coopers, Datsun 1600s and similar.

Now the new engines — the 1300, 1600 and 1600 GT all get the new cross-flow design — have put back the snap into what has always been one of the most favored of young men's cars. Our fondest memory of a Cortina GT was the 1964 two-door which the Editor bought after it almost won the Armstrong 500 in the hands of the Geoghegan brothers. This would spin to 7000 rpm like a whip crack, just top 100 mph, and put away a quarter in the high 18s. It was, of course, finely tuned and "blueprinted" and had that indefinite air and "zing" common to all well-balanced engines. The new car has the same feel to it. Red-lined from 6000 to 7000 rpm, it will reach 7000 effortlessly, although 6200 is a sensible limit both for safety and performance curves. And the current car has far better seats and equipment, as well as better looks, even though the old car had a pleasantly lean and hungry air about it.

Why the new engines? We fully described the new Heron head system in WHEELS a few months back, but as a brief reminder let us just say that by placing inlet and exhaust porting on opposite sides of the head the engineers can get a much straighter gas flow and thus more efficient combustion. But more importantly, by adopting this "pent-roof" shape for the combustion chamber they can use proportionately larger valves and thus get much better breathing. To this, Ford engineers went a stage further and added the idea first tried in their V4 engines — that of recessing the chamber itself in the crown of the piston, enabling the spark plug to be placed more centrally and thus promoting even better combustion.

You can pick the GT from the other Cortinas by different badges, a partly blacked-out grille, no thin chromed "fin" atop the rear wing, but a narrow chromed lip over the wheel arches and along the door sills. It also has slightly fatter tyres — 5.60 against 5.20, but apart from that the interior, with its good bucket seats divided by a central console bin and the extra instrumentation and carpeting, is the only guide. The mechanicals are the same, except for boosted engine power, slightly bigger and thicker discs and drums, a 3.9 final drive against the 4.44 of the 1300 and 5.125 of the ordinary 1600, closer gear ratios and extra weight — 2023 lb kerb against 1913 for the 220 1300 and 1992 for the 440 1600.

The new engine has brought a considerable power lift — up from 83.5 bhp at 5200 rpm to 93 at 5400, with torque going up from 97 ft/lb at 3600 to 101.5 at the same point. The better breathing is immediately obvious in the lift in the rpm peak, but we are a little surprised at the relatively smaller lift in torque, particularly as the increase in volume was done by stroking from 2.867 in. to 3.056. This would have the effect of raising piston speeds but also producing more torque lower down in the range. However, the result is definitely a flatter torque curve, and the car reflects this.

It is much nicer to potter with — as if anybody should potter in

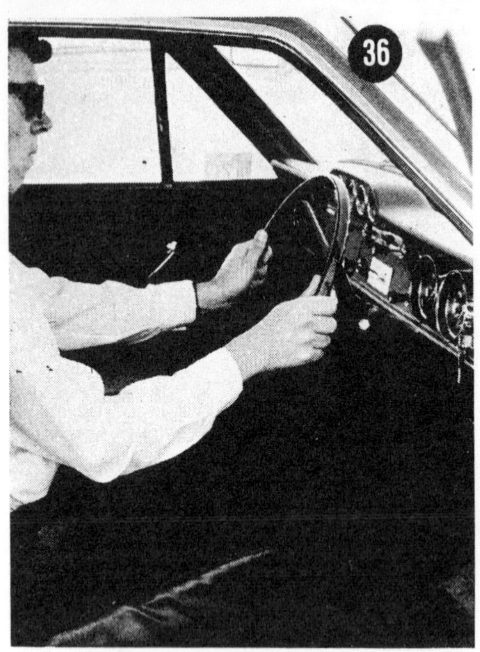

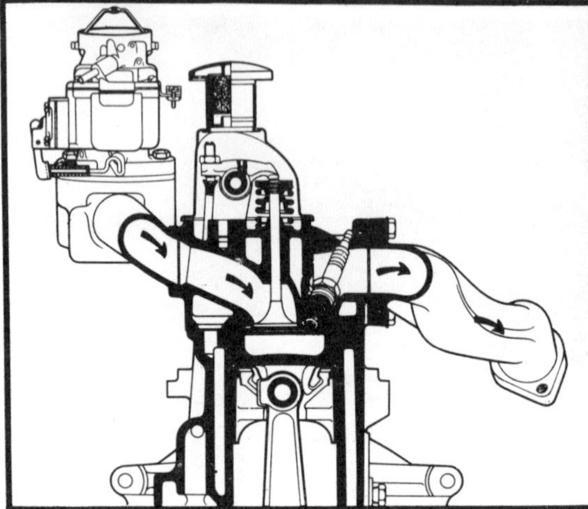

Seats look ordinary, but are actually very, very good. This is interior of previous 1500, with black wheel and without console badge.

Schematic of new head shows how central placement of plug and larger valve sizing is encouraged by cross-flow design and flat head.

a GT. But the effective reservoir of power starts below 2500 rpm and extends to just over 6000 and when you match this to a well-selected seat of close ratios — closer, that is, than in a more pedestrian car and certainly without the agonisingly low second gear of the 1964 car — you get the feeling you get in all essentially good open road cars of never actually being in the wrong gear.

Acceleration from standstill to 70 mph of 17.5 sec is good for a 1600 cc pushrod engine, and the top gear acceleration figures are also good, reflecting the fact that by lengthening the stroke Ford has removed much of the bogey of poor low-end torque with a very over-square engine — about the only penalty of that design.

Braking is as always — light pressures and deadly secure stopping, coupled with excellently progressive "feel". The ride is good over most surfaces, with some choppiness on rutted roads, but the car always tracks straight and true with only a touch of the traditional Cortina "dodging about" when running in a straight line with some side-winds. The car understeers slightly and consistently, but can be "set up" for corners in a big way beforehand. It has the same excellent trait of good rear end stickability under full power, and it is hard to make it lift an inside wheel on hairpins. On gravel roads it skates around a little, but stays safe and stable all the time.

Inside the car the nose level is low, and there is little reflected road noise. In fact, the whole structure feels very taut and well-insulated. The steering is light and neat, transmitting just enough road information. It is a very quiet car with all the glass up and the Aeroflow ventilation working, but our testers agreed that the volume of air being put through with the new butterfly valve was not even half that of the old system of a simple flap. We found we could not get the air inside the car exchanged quickly enough to provide ventilation in a 90-degree Sydney summer; a window just had to be opened. And the quarter vents are fixed.

Unfortunately Ford has put all its shekels into performance and spent little on finish inside. The interior reeks of the 220, which is a severe comment, for the 220 has by far the poorest finish of any car in the Ford range. Sure, there are nice hooked nylonised carpets in the GT, and very good black sponge-backed pvc, but the other materials, including that awful white steering wheel, are straight from the cheaper cars. Also, there is little finishing-off — the headlining just ends where it ends, alloy door sill trims have raw edges, the nylon-bushed lid to the central locker doesn't close properly, and the windscreen washer button is one of those pitiable vacuum spit-and-promise affairs.

There is full instrumentation, but it is not much use, as the gauges for fuel, oil pressure, temperature and amps are mounted in a nacelle atop the centre of the dash, where they are hard to read at all, particularly at night. Ergonomics has given way to economics here, for the nacelle simply fits on to a preformed recess in the standard dash panel and is covered by a strip of padding — saves forming another facia for the GT. The speedometer and tachometer, however, are very well-placed and very steady. Our photos, incidentally, show the air vent controls each side of the heater-demister slides. This was an earlier car that had not got the butterfly valve system in the dash vents.

But that's where the maltreatment of the driver ends. He gets seats which don't look comfortable but which are in fact excellent, with good support everywhere, even against the high sideways loadings this car can generate in skilled hands. Unfortunately they moved on one of those dreadful U-slides in which the squab becomes more vertical the further the seat goes back, but we found a compromise position which still gave us good arm length at the wheel. They are good long-distance seats.

Continued on page 12

TECHNICAL DETAILS

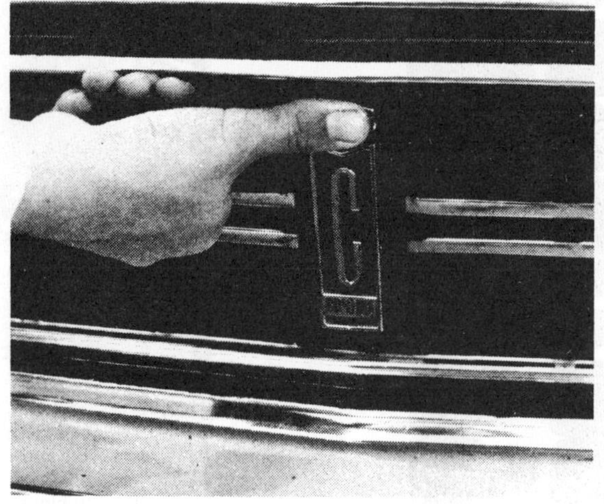

MAKE	Ford	**PRICE**	$2530
MODEL	Cortina 1600 GT	**COLOR**	British racing green
BODY TYPE	4-door sedan	**MILEAGE, START**	1388
		MILEAGE, FINISH	1565
OPTIONS	radio	**WEIGHT**	$17\frac{3}{4}$ cwt

TEST CONDITIONS:

Weather: fine, hot. Surface: hot mix bitumen. Load: two persons. Fuel: Fleetwing Supreme

PERFORMANCE

Piston speed at max bhp .... 2580 ft/min
Top gear mph per 1000 rpm 17.1
Engine rpm at max speed 5600
Engine rpm at cruising speed .... 4700 at 80 mph
Lbs (laden) per gross bhp (power to weight) 22.2

MAXIMUM SPEEDS:

Fastest run .... 95.7 mph
Average of all runs .... 95.0 mph
Speedometer indication fastest run .... 98 mph
In gears: 1st 35 mph; 2nd 50 mph; 3rd 70 mph at 6000 rpm.

ACCELERATION:

(through gears)

0-30 mph	4.2 secs
0-40 mph	6.9 secs
0-50 mph	9.5 secs
0-60 mph	13.1 secs
0-70 mph	17.5 secs
0-80 mph	25.0 secs

	3rd gear	4th gear
20-40 mph	6.7 secs	11.3 secs
30-50 mph	6.4 secs	10.2 secs
40-60 mph	6.8 secs	10.1 secs

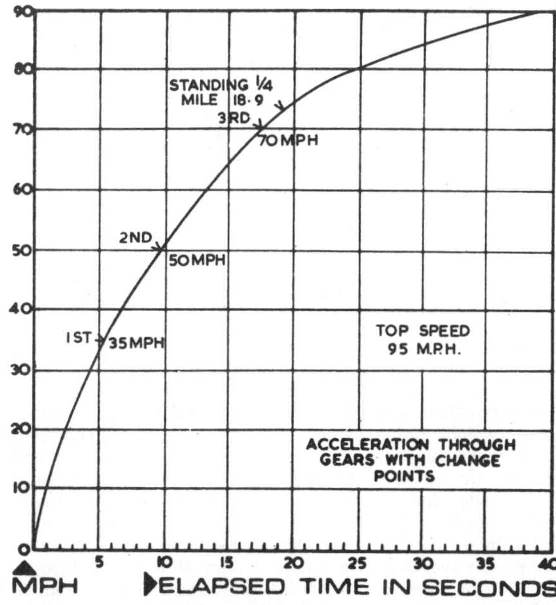

STANDING QUARTER MILE:

Fastest run .... 18.8 secs
Average of all runs .... 18.9 secs

SPEEDOMETER ERROR:

Indicated (mph):	30.0	40.0	50.0	60.0	70.0
Actual (mph):	30.0	39.0	48.0	58.5	69.0

SPECIFICATIONS

ENGINE:

Cylinders .... four in line
Bore and stroke .... 80.9 mm by 70.62 mm
Cubic capacity .... 1599 cc
Compression ratio .... 9.0 to 1
Valves .... overhead, pushrod
Carburettor .... Weber 2 barrel compound
Power at rpm .... 88 bhp at 5400
Torque at rpm .... 96 ft/lb at 3600

TRANSMISSION:

Type .... four speed all syncro
Clutch .... 7.54 in. dia diaphragm
Gear lever location .... central, in console
Overall ratio: 1st 11.591; 2nd 7.839; 3rd 5.448; 4th 3.900; Final drive 3.900 to 1.

CHASSIS and RUNNING GEAR:

Construction .... unitary
Suspension, front .... Macpherson strut, coils
Suspension, rear .... leaf springs
Shock absorbers .... telescopic Armstrong
Steering type .... recirculating ball
Turns 1 to 1 .... $4\frac{1}{2}$
Turning circle .... 30 ft
Steering wheel diameter .... $15\frac{1}{2}$ in.
Brakes, type .... disc front, drum rear
Dimensions .... 9.625 in. dia disc, 9.0 in. dia drum

DIMENSIONS:

Wheelbase .... 8 ft 2 in.
Track, front .... 4 ft $4\frac{1}{2}$ in.
Track rear .... 4 ft 3 in.
Length .... 14 ft
Height .... 4 ft $8\frac{1}{2}$ in.
Width .... 5 ft $4\frac{3}{4}$ in.
Fuel tank capacity .... 10 gals
Tyres, size 165-13
Make on test car .... Olympic GT radial
Ground clearance, registered .... 6.6 in.

CONTINENTAL CASEBOOK

1600E

WE CANNOT BE CALLED THE most enthusiastic fans of Ford's current Executive idea, not least because the word itself with its social-climbing admass overtones is so very repulsive. But fundamentally we can indeed see the point. The late Laurence Pomeroy, of this address, was one of the first to point out what a hideous blunder the men of Dagenham had made in failing to build up a 'quality image' that buyers of their cheap and nasty Populars could aspire to—an image that would attract customers whom the very existence of the Populars and their ilk would otherwise frighten away. Pom wanted Ford to buy Alvis, which he rightly saw as the last of the great independent marque names, but Rover got it instead and promptly decided to kill it after producing a prototype or two. So by the time he had succeeded, through his great friend Harley Copp (now, alas, returned to Detroit having opened the eyes of all of us to the true meaning of body engineering), in convincing management of the wisdom of his plan the only way out was to create a new prestige name from scratch. Rightly or wrongly, the choice fell on Executive—and got off to a bad start with abortive attempts to make something of those resounding commercial clangers, the Zodiacs III and IV.

But with the arrival of the Corsair 2000E, just in time to save Ford from another painful experience, things had begun to look up. And the 1600E Cortina—latest of the line, though soon to be followed presumably by a variant on the Escort—is almost good enough to change our minds about the idea as a whole.

The difference with these latest cars is that they are not just a pretty face, if pretty happens to be your word for a thick plank of varnished wood and lots of matt black vinyl. They are subtle combinations of wanted features which exist singly elsewhere, as in the best of the enormously successful American 'sports-personal' cars.

The 1600E, for example, offers the practicality of four-door saloon bodywork and the undeniable lure of a fully instrumented, adjustable, bucket-seated treewood-and-carpet interior together with the simple thrust of the Cortina GT pushrod engine in crossflow form and the superior roadholding that comes with ultra-low, ultra-hard Lotus-Cortina suspension, wide wheels and radial tyres—all topped off with a lacing of extrovert gingerbread, right down to chromed hubs, metallic paint and a thin gold line.

Singly, these things might even entail disadvantages that are eliminated automatically in the E. Cortina GTs, for example, are notorious for harshness and general lack of insulation, whereas the 1600E carries more anti-NVH ingredients than any other model in the range without any noticeable loss of performance. The latest crossflow engine can be much rougher and noisier than its predecessor in more mundane applications, too, but the E's thick carpeting and other applied insulation helps bring the noise level back within bounds. The philosophy obviously goes this way: that the very man whose taste runs to drawing-room superficialities in his car is just the type who might need to get a move on, and for whom the old BMC idea of merely adding a half-hundredweight of walnut and a compensatory SU is patently inadequate.

But 10 days spent with a well-used 1600E both in London and on the Continent showed us that there are formidable gaps in the Ford formula. For example, wide wheels and lots of negative camber might do wonders for Mr Executive's ego but they virtually exclude Mrs E from the driving seat. That padded-rim wheel connects to steering *far* too heavy for any female biceps we know, and the sporting geometry can send back some formidable road shocks as well as removing much of the normal Cortina's inherent self-centring action. The GT engine's camminess and lightweight, rather delicate clutch combine, too, to make the car rather tender in the mouth for any but an experienced operator to handle with equanimity.

Built-in compliance allows Ford to fit radials to almost any of their models without tears, but the drastically shortened wheel travel that comes with the 1600E's 'handling package' brings problems in its wake. Even baby's-bottom British byways set the unsympathetic seat-springs dancing, so that passengers (who have no wheel to comfort them) are bounced about unmercifully—especially in the back. And on French roads! We are tempted to say that the E gives its occupants an even rougher time than the average sports car, but reflection on the agricultural nature of most current contenders for that once-coveted title causes us to observe merely that it is a great deal worse than any European saloon when it comes to comfort and wellnigh intolerable for rear seat passengers irrespective of tyre pressures or load. A seasoned traveller who accompanied us to Switzerland and back was sick in this car for the first time in his life, and for the driver the effect of all that jiggling was simply exhausting. Executives, it seems, are a purely British phenomenon, and must remain so.

There is another unwelcome intruder: wind noise. Ford are at last responding to constant nagging from us and from customers about this, but so far the results (to take the form initially of tubular door seals, as on Renault's R16) have not reached production. The Cortina's separate window frames have always been prone to bulging, so that air can leak past the top corners and cause a most unwelcome roar. The faster you go the worse is the noise, and the 1600E is almost in the genuine 100mph class. Above 80 things are so bad that the radio is inaudible and conversation becomes too much effort—hardly the thing after a busy day in the office.

The positive side? Well, almost everything else works out jolly well. With the exception of that heaviness at parking speeds, all of the controls are a delight—especially the gearbox, which gets better every time we try it and comes of course in this application with the GT's higher second gear (but also, alas, with a maddening lever vibration such as Lotuses and Mini-Coopers used to suffer from). The seats with their adjustable backrests are among the best in this price class for long trips, with just the right degree of firmness and location; our only suggestion is for a trifle more rearward adjustment. Instrumentation is splendid, and the minor controls are well planned. Heating, superb; but the wipers travel *far* too slowly and the headlamps are inadequate.

Perhaps the car's most endearing attribute is its one-piece handling, an important ingredient of which is the engine's adequate power. Understeer is the prevailing trait, but a good bootful of throttle will soon convert it to the other thing. Excessive roll and tyre squeal are both absent for once, although the rear axle's rudimentary location precludes the taking of liberties on bumpy bends. Wet weather roadholding (Dunlop SPs, the first time we've had 'em on a Ford) is exemplary and braking power reassuring in most circumstances, with minimal effort and no premature locking, though steady punishment in the mountains did bring us down to less than 50percent efficiency at one stage. Recovery: almost immediate.

All told, the 1600E is a clever mixture but still, inevitably, a bit of a curate's egg. At £1029 tax paid and with all extras except a radio (ours, the usual Plessy printed circuit affair, was simply dreadful) it represents value for money as an only car for the essentially rather selfish man who likes to do all the driving himself and to press on regardless, valuing all-round performance more than comfort. On continental roads it is not a proposition as a family car at all, though judged as a two-seater it is no worse than most and can keep up with all but a few. *DB*

GROUP TEST

'Motor's' test team go driving in convoy to try groups of competitive cars under identical conditions

No. 5: £1,000 saloons

● Ford Cortina 1600 GT ● Ford Corsair 2000 de luxe ● Austin 1800 de luxe ● Hillman Hunter Mk. II ● Vauxhall Victor 2000 ● Renault 16 GL

The cars

Family men with two or more children looking for comfortable but lively transport and middle-level company executives are the sort of people at whom this group of cars is aimed—4/5-seat saloons costing just under £1,000.

Ford Cortina 1600 GT four-door £964. A front-engined/live rear axle car which lies roughly in the middle of Ford's extensive and recently restyled Cortina range using an advanced new line of Heron-headed engines.

Ford Corsair 2000 de Luxe £968. Larger externally and fitted with a powerful V-4 front engine; independent front suspension and live rear axle. Is the same mechanically as the more luxurious 2000E model.

Austin 1800 de Luxe £999. BMC's biggest front-wheel-drive car with all-independent suspension and prodigious internal space—rear seat legroom particularly. Now a fully sorted car, but one whose reputation has suffered badly through early development faults. (Tested in "Mk. I" form).

Hillman Hunter Mk. II £1,012. A comfortable and well-equipped four-seater saloon with independent front suspension and a live rear axle now produced in an improved Mk. II version.

Vauxhall Victor 2000 £973. Vauxhall's new medium-sized saloon with the General Motors family styling and a belt-driven overhead camshaft 2-litre engine. Front engined with independent front suspension and a live rear axle.

Renault 16 GL £998. Front-wheel-drive car with a refined four-cylinder engine and body styling which is a compromise between the saloon and the estate car layouts.

AMPLE space at a moderate price is the feature common to the cars in this group; a family and the ability to raise—or persuade their employers to raise—nearly £1,000 are probably the only characteristics shared by the people who buy them. At one extreme is the buyer who takes no pleasure in driving, cares nothing for the sporting aspects of motoring and wishes only for a machine that will transport his brood in comfort with enough power to climb hills easily and maintain what seems to him to be a reasonable pace. At the other extreme is the sporting driver, interested in performance handling and roadholding—the driver who likes to press on. Given a little more money he would probably choose one of the cars in last week's group—the BMW 1600 or the Rover 2000, say—but the cheapest of these demands a further £250 or more and so is out of reach.

The cars themselves vary almost as much in mechanical specification and character, but it is a measure of recent advances in design that many of them succeed in satisfying the conflicting demands of different kinds of customer. A few years ago cars of this type tended to be wallowy devices taking up a fair amount of road-space without providing a proportionate amount of room inside. Now, however, cars like the BMC 1800 and the Renault 16 vie with each other in the amount of interior space provided and the ingenuity with which it can be arranged. At the same time both these cars have very good handling and roadholding as does the Victor 2000. Similarly, different virtues are to be found in the other cars that make up the group.

Performance

The performance of the cars falls roughly into two groups. In the first—composed of the Corsair (fastest of all the cars tested), the Cortina and the Victor—maximum speed approaches 100 m.p.h. (97.9-94.4 m.p.h.) with acceleration to match. The Austin 1800, the Hillman Hunter and the Renault 16 form the second group and are rather slower having maximum speeds around 90 m.p.h. (92.8-85.9 m.p.h.). As the performance charts show, the Hunter, surprisingly, has the best top-gear pull, but this is mainly because the figures quoted refer to a car fitted (like our test car) with overdrive which has a low direct top gear. The Victor 2000 and Corsair 2000 are almost as good, while the slowest car in the group—hampered by a relatively small capacity (1,470 c.c.) and a high top gear—is the Renault 16. For motorway cruising however, this high top gear proved an advantage, as did the overdrive of the Hunter since these two cars were judged to be the least fussy at 70 m.p.h. along

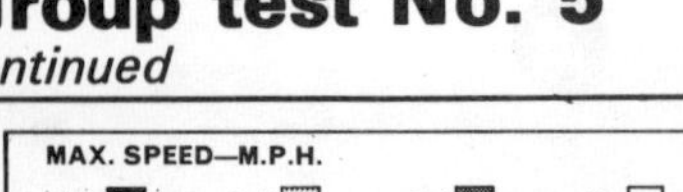

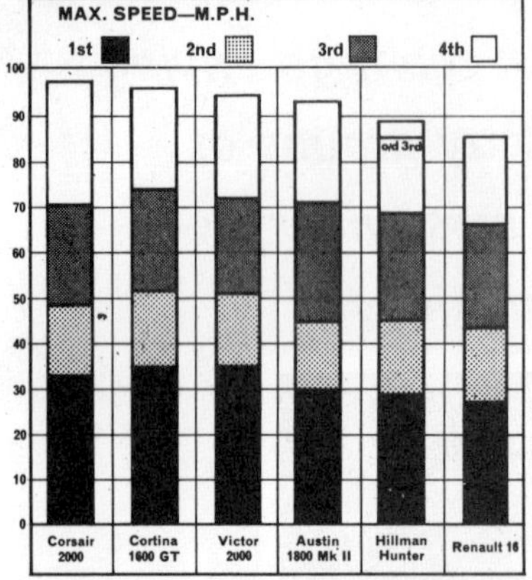

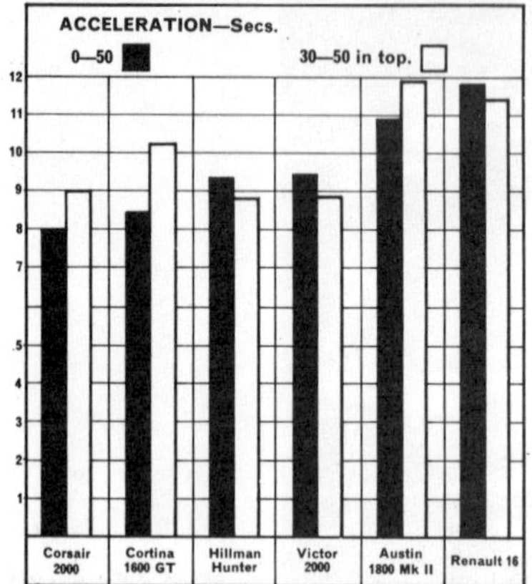

All performance figures are taken from the original road tests (as listed on page 176): those quoted for the Ford Cortina 1600 GT and the Corsair 2000 de luxe are taken from the tests of the mechanically identical 1600E and the 2000E, while those quoted for the Hillman Hunter are taken from a test of the "MK. I" version.

with the Austin 1800. At such speeds the Cortina and Victor were spoilt by wind noise as was the Corsair to a lesser extent.

Both the Austin and Renault engines were judged about equal for smoothness, and both these units had the equally important quality of willingness and the ability to take high revs without fuss. Although the Cortina was much smoother than some other examples that we have tested recently, it could not be placed in the front ranks even though most felt it to be better than the Victor and the Hunter which were generally considered to be about the same in smoothness. Worst by a considerable margin was the Ford Corsair which had a rough, boomy engine note, although this was to some extent compensated for by lively performance, good low speed torque and outstanding flexibility.

Economy

On our 500-mile trip to Wales fuel consumption varied quite widely from the 20.5 m.p.g. of the Victor and the 20.8 m.p.g. of the Austin to the 27.2 m.p.g. of the Renault and the 27.4 m.p.g. of the Hunter. Since both the Hunter and the Renault have Group 3 insurance rather than Group 4 which is required for all the other cars (except the Austin 1800 which also needs Group 3) these two cars should be significantly cheaper to run than the rest. Rather different fuel consumptions were

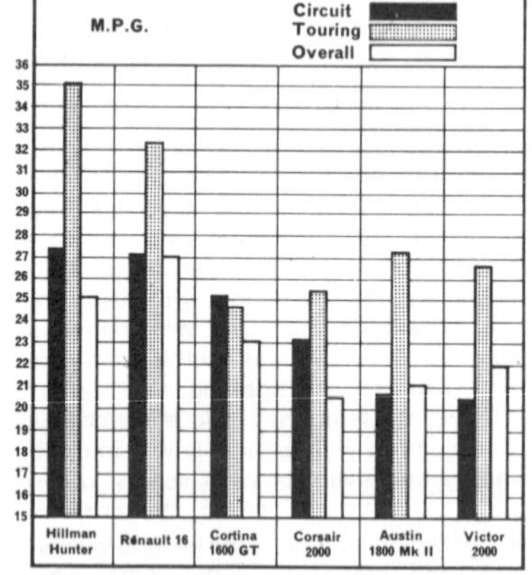

obtained during our original road tests, but the Hunter and the Renault are confirmed as being the least thirsty.

Transmission

All the cars were fitted with four-speed synchromesh gearboxes, all but the Renault having floor changes, while the Hunter was fitted with the optional overdrive. Both the Cortina and Hunter had excellent changes: light, precise, with good synchromesh, and the Corsair was almost as good, although its action was a little spoilt by a long-travel clutch. The Victor change, however, was a little notchy and obstructive. Although few of our testers have much enthusiasm for column gearchanges it is a measure of the excellence of the mechanism fitted to the Renault—and indictment of that of the Austin—that we judged the Renault mechanism the better: it was light and precise in action, and it allowed quick downward changes into second when these were occasionally required for hairpins. This is rarely easy with column changes. The Austin change, on the other hand, although quite pleasant to use when not in a hurry, had a rather vague and notchy movement and was often obstructive in first and second.

Most of the cars had well spaced ratios, while the combination of low direct gears and high overdrive on the Hunter is an excellent one. Although top gear was high for motorway cruising on the Renault, this was rarely a disadvantage since nearly 70 m.p.h. was possible in third. Transmission noise was moderate on the majority of the cars, the exceptions being the Victor which had noisy indirects and the 1800 of which the transfer gears chattered badly when idling.

Handling and roadholding

There was a division of opinion amongst our testers as to which cars had the best steering. The first and most powerful faction considered that the feel and precision of the 1800 steering more than compensated for its low gearing, that the R16 mechanism was responsive

Although the Renault 16 rolls a lot, its handling has been greatly improved since it was first introduced

Lined up for the start—at Portmeirion, known to television viewers (but not in this view) as the scene of the series "The Prisoner". The Hunter leads the way; seen here in its square-headlamped Mk. II form.

despite being rubbery, and that the lightness and accuracy of the Victor's steering outweighed its low gearing.

A minority opposition felt that the Cortina and the Hunter were the best, with light, reasonably precise and direct steering giving good feel of the road, while the Corsair was considered to be a little less accurate. The 1800 steering was judged to be just as precise and to give good feel but was marked down through being low-geared. Excessive rubberiness was deemed to be the fault of the Renault steering, and the Victor was thought to be the worst because it was light and dead yet suffered from kickback.

The live axled cars held the road quite well so long as the surface remained smooth but their tails tended to hop and slide on bumps, except for the Victor which was quite good on both rough and smooth roads. With all-independent suspension the two front-wheel drive cars held the road just as well as their cart-sprung rivals on smooth surfaces but did much better in the rough: although they occasionally lifted a front wheel on humps, this rarely made the car change direction. For these reasons the 1800 was considered to have the best road-holding, followed by the Victor, the Renault 16, the Corsair, the Cortina and the Hunter, in that

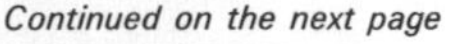
Continued on the next page

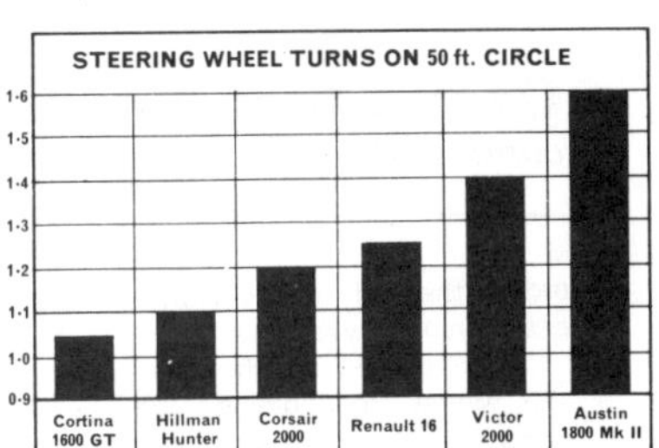

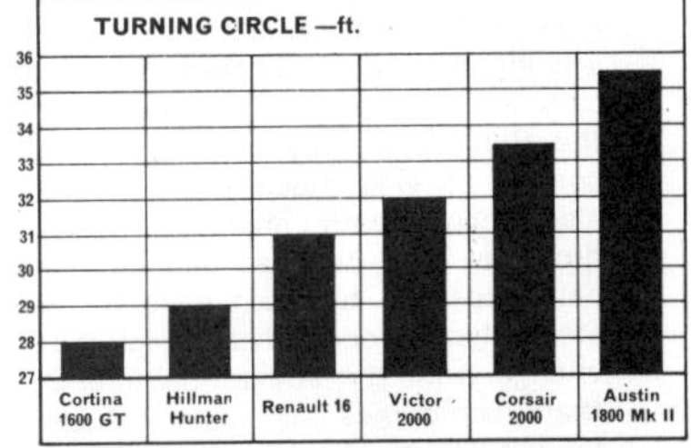

On twisty roads like this the Cortina's roadholding was only fair, but many of our testers liked its responsive handling characteristics.

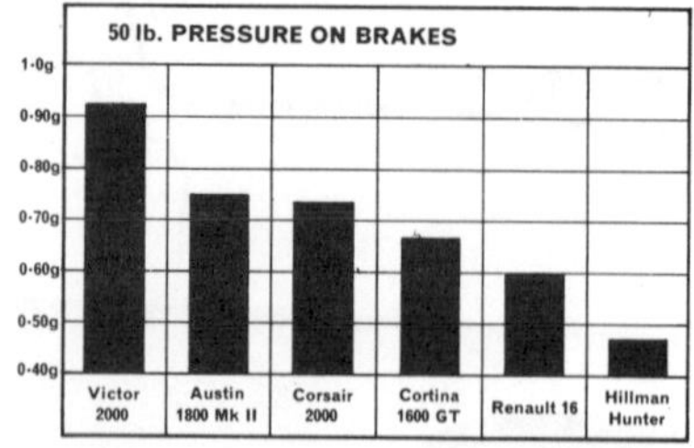

Brakes

Half the cars had light, progressive brakes with good feel characteristics, but the Austin 1800 servo was a little obtrusive and unprogressive, the Renault's brakes were rather heavier than the others while the Hunter's were slightly spongy.

Comfort

Adjustable for rake but needing a larger range of fore-and-aft adjustment, the seats of the Renault 16 were nevertheless found to be the most comfortable: the small padded odds-and-ends box between them improved lateral support. Almost as good—except in this particular quality—were the reclining (optional) seats of the 1800, while the Hunter's (reclining as standard) were also

Group test No. 5
continued

order, the tail of the Hunter hopping very easily on bumpy roads.

Some drivers felt that the car with almost the worst roadholding—the Cortina—had good handling because of its responsive behaviour and the controllable nature of its progressive final oversteer. The Hunter behaved in a similar way but the oversteer was generally promoted by bumpy surfaces rather than hard cornering or the application of power while the Corsair was rather wallowy and suffered from excessive understeer. It was felt that the good handling of the Victor would be more enjoyable if its steering was much higher geared—which it easily could be without calling for undue effort when parking. With a fair amount of roll, considerable understeer and very good adhesion the 1800 was less interesting to drive than most of the other cars. At low speeds roll and tyre scrub made the Renault tedious to drive, but on faster corners it was much better. Handling has also been greatly improved since we last drove an R16 (we had one on our staff), it being possible to attain a gentle final oversteer when cornering hard which is a welcome change from the straight-on final understeer that sometimes characterizes the more powerful front-wheel drive cars.

Lively performance and good flexibility were two of the Corsair's good points.

The Victor had extremely good roadholding but driving pleasure was a little spoilt for some by the low-geared steering.

considered to be good. Few drivers liked the Ford seats because of too small an angle between seat and backrest, because the seat tilts forward as it is moved back and because there is insufficient fore-and-aft adjustment, particularly in the Corsair. Opinions varied about the Victor seats, some complaining about the lack of lateral support and finding the stiffness of the lumbar support objectionable.

For smoothness of ride the Renault was unanimously put to the top of the class, and the general lack of road and wind noise in this car helped to reduce fatigue when driving it. Nearly everyone felt that the 1800's Hydrolastic suspension did almost as good a job, the virtual elimination of pitch being most effective, but a few found the vertical movements excited by certain types of undulation most uncomfortable. The live-axled cars tended to pitch on rough surfaces, the Victor being the best with the Hunter and Cortina the worst.

KOF 120E

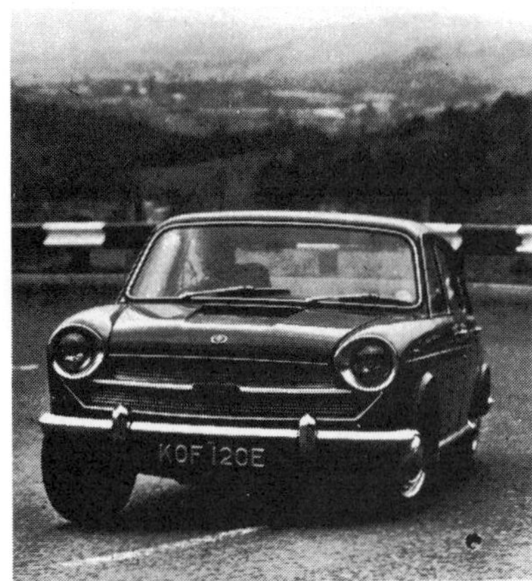

As the rally experts have shown, the 1800 can be thrown about like a Mini.

Accommodation

When it comes to packaging people and their luggage the BMC transverse engine and front wheel drive formula certainly pays off, for the Austin 1800 at 13ft. 8¼in. is over a foot shorter than the longest car (the Corsair) yet has the biggest boot and by far the most spacious interior. The rear seat legroom in particular is so large that if at some time in the future the car is ever restyled the designers might well consider reducing the length of the interior space an inch or two to make the boot still bigger. It is also possible to seat three large men in the back, and although the

Continued on the next page

BOOT CAPACITY—cu. ft.
13
12
11
10
9
8
Austin 1800 Mk II
Victor 2000
Corsair 2000
Renault 16
Cortina 1600 GT
Hillman Hunter

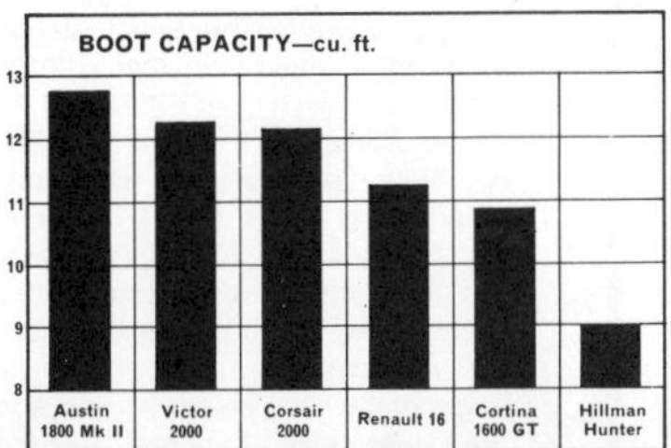

Vital statistics

OVERALL WIDTH 5'-7½"
REAR TRACK 4'-7½"
FRONT TRACK 4'-8¼"
GROUND CLEARANCES
LOWEST POINT (UNDER ENGINE) 6½"
UNDER FRONT SUSPENSION 6½"
UNDER EXHAUST 7"
SCREEN FRAME TO FLOOR 42"
FLOOR TO ROOF 45"
4'-7½" UNLADEN HEIGHT
BOTTOM OF DOOR TO GROUND 14"
8'-10"
13'-10¼"
HEIGHT OF MALE FIGURE 5'-10" APPROX.
HEIGHT OF FEMALE FIGURE 5'-7" APPROX.
SEAT MEASUREMENTS TAKEN WITH SEATS COMPRESSED

OVERALL WIDTH 5'-5"
REAR TRACK 4'-3"
FRONT TRACK 4'-5"
GROUND CLEARANCES
LOWEST POINT (UNDER FRONT SUSPENSION) 7½"
UNDER EXHAUST 7½"
UNDER ENGINE 8"
SCREEN FRAME TO FLOOR 43"
FLOOR TO ROOF 46½"
4'-9" UNLADEN HEIGHT
BOTTOM OF DOOR TO GROUND 14"
NEARSIDE 8'-11" OFFSIDE 8'-8½"
13'-10¾"
HEIGHT OF MALE FIGURE 5'-10" APPROX.
HEIGHT OF FEMALE FIGURE 5'-7" APPROX.
SEAT MEASUREMENTS TAKEN WITH SEATS COMPRESSED

OVERALL WIDTH 5'-5½"
REAR TRACK 4'-4½"
FRONT TRACK 4'-5½"
GROUND CLEARANCES
LOWEST POINT (UNDER FRONT SUSPENSION) 5½"
UNDER EXHAUST SYSTEM 5½"
UNDER ENGINE 7"
SCREEN FRAME TO FLOOR 41"
FLOOR TO ROOF 45"
4'-8" UNLADEN HEIGHT
BOTTOM OF DOOR TO GROUND 11½"
8'-2"
14'-0"
HEIGHT OF MALE FIGURE 5'-10" APPROX.
HEIGHT OF FEMALE FIGURE 5'-7" APPROX.
SEAT MEASUREMENTS TAKEN WITH SEATS COMPRESSED

OVERALL WIDTH 5'-4"
REAR TRACK 4'-4"
FRONT TRACK 4'-4½"
GROUND CLEARANCES
LOWEST POINT (UNDER FRONT SUSPENSION) 6½"
UNDER EXHAUST 8½"
UNDER ENGINE 8½"
SCREEN FRAME TO FLOOR 40½"
FLOOR TO ROOF 44½"
4'-8" UNLADEN HEIGHT
BOTTOM OF DOOR TO GROUND 13½"
8'-3"
14'-1½"
HEIGHT OF MALE FIGURE 5'-10" APPROX.
HEIGHT OF FEMALE FIGURE 5'-7" APPROX.
SEAT MEASUREMENTS TAKEN WITH SEATS COMPRESSED

OVERALL WIDTH 5'-7½"
FRONT TRACK 4'-6½"
REAR TRACK 4'-6½"
GROUND CLEARANCES
LOWEST POINT (UNDER FRONT SUSPENSION) 5½"
UNDER EXHAUST 6½"
UNDER ENGINE 6½"
SCREEN FRAME TO FLOOR 39"
FLOOR TO ROOF 43½"
4'-6" UNLADEN HEIGHT
BOTTOM OF DOOR TO GROUND 11½"
8'-6"
14'-8½"
HEIGHT OF MALE FIGURE 5'-10" APPROX.
HEIGHT OF FEMALE FIGURE 5'-7" APPROX.
SEAT MEASUREMENTS TAKEN WITH SEATS COMPRESSED

OVERALL WIDTH 5'-3½"
REAR TRACK 4'-2½"
FRONT TRACK 4'-2½"
GROUND CLEARANCES
LOWEST POINT (UNDER FRONT SUSPENSION) 6½"
UNDER EXHAUST 8"
UNDER ENGINE 8½"
SCREEN FRAME TO FLOOR 40"
FLOOR TO ROOF 45¼"
4'-9" UNLADEN HEIGHT
BOTTOM OF DOOR TO GROUND 12½"
8'-5"
14'-11¾"
HEIGHT OF MALE FIGURE 5'-10" APPROX.
HEIGHT OF FEMALE FIGURE 5'-7" APPROX.
SEAT MEASUREMENTS TAKEN WITH SEATS COMPRESSED

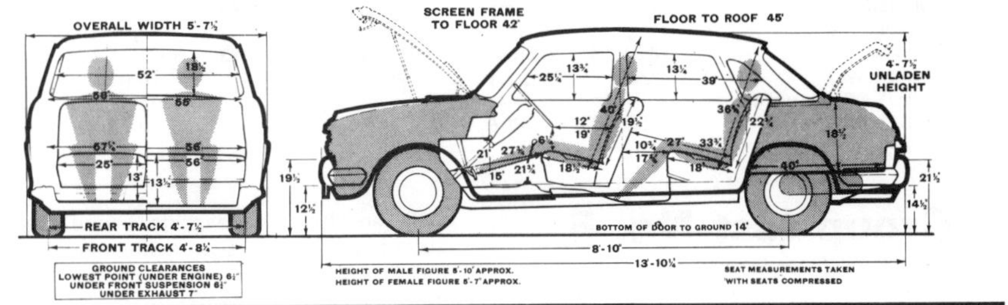

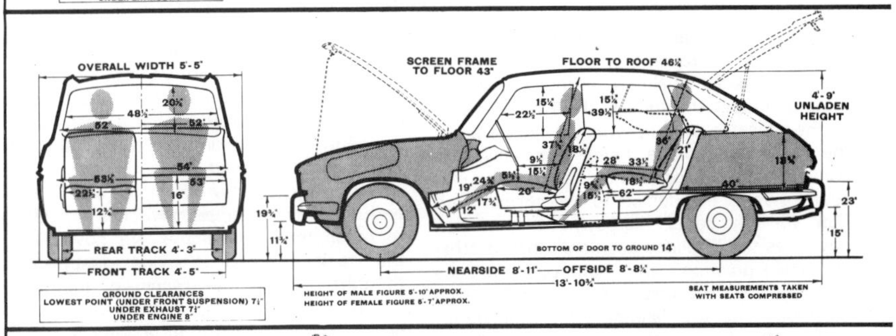

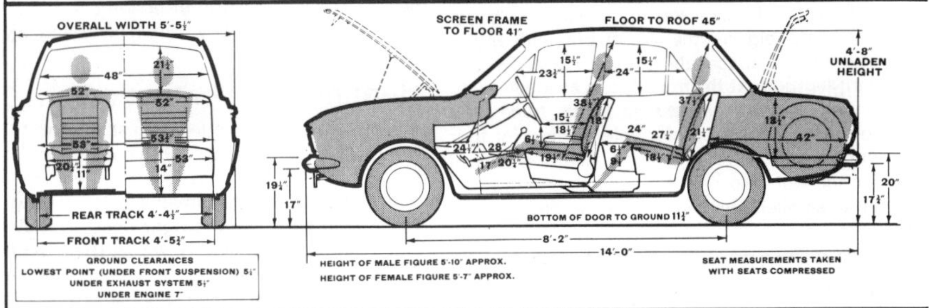

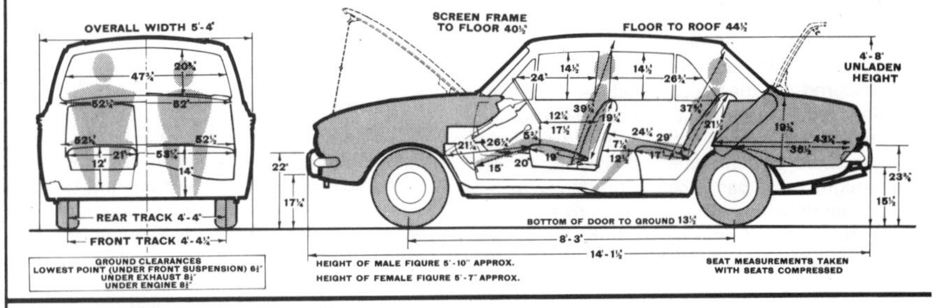

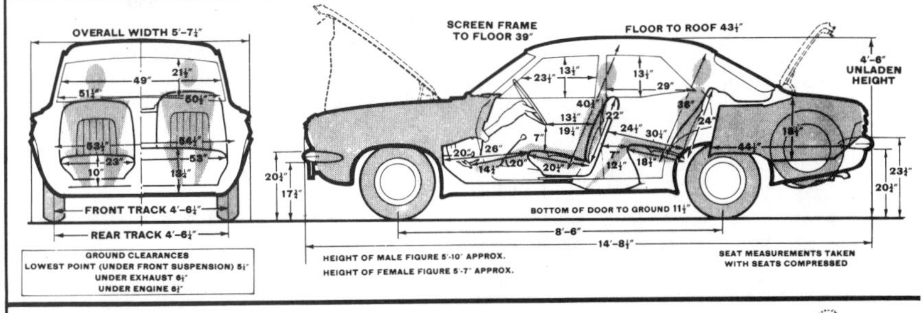

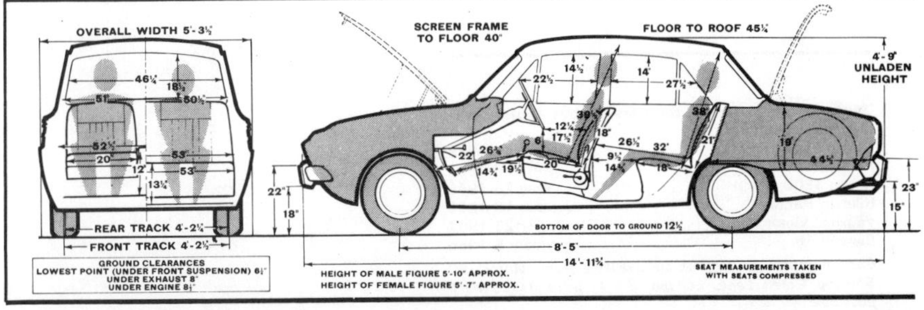

Specifications	Austin 1800 Mk II	Corsair 2000	Renault 16	Cortina 1600 GT	Hillman Hunter	Victor 2000
Cylinders	4	V-4	4	4	4	4
Capacity	1,798	1,996	1,470	1,599	1,724	1,975
Brakes	discs/drums	discs/drums	discs/drums	discs/drums	discs/drums	discs/drums
Service intervals	6,000	5,000	3,000	6,000	5,000	6,000
Fuel grade	4-star	4-star	4-star	4-star	4-star	4-star
Insurance rating	3	4	3	4	3	4
Tyre size	165 × 14	165 × 13.	145 × 355	165 × 13	5.60 × 13	6.2 × 13
Weight (cwt.)	22.7	21.1	22.6	18.9	18.0	21.1
m.p.h./1,000 in o/d top	—	—	—	—	21.7	—
m.p.h./1,000 in top	18.1	17.6	17.2	17.2	17.4	16.5

Group test No. 5
continued

1800 is wider than most of the other cars, it is fractionally narrower than the Victor.

The Renault has almost as much room but must be judged the equal of the 1800 because of the versatile way in which it can be used, the basic advantage being the ability to convert to estate car form by folding the rear seat forward. In addition to the normal position for its seats, its owner's handbook describes the "holiday position", the "mother and child position", the "rally position", the "travel bed position", the "bulky cargo position" and the "shooting brake position".

Both the Corsair and the Victor have boots nearly as big as the 1800 but with rather less interior space, the Corsair having insufficient legroom at the front due to an inadequate range of seat adjustment. The two cars, however, provide comfortable seating for four with room for five at a pinch. The Cortina and Hunter will also seat four in comfort but are rather smaller cars with a little less boot space.

Instruments and switches

Perhaps the best set of instruments and minor controls are those fitted to the Cortina, the only criticism being that the group of instruments in the raised nacelle at the centre of the facia are for some drivers partly obscured by the steering wheel. The Corsair also had a pleasant-looking set of instruments, but the important lights and wiper switches are small and mounted almost as an afterthought on a panel below the facia. An even bigger defect has been built into the control layout of the Hunter in which the lights switch is immediately behind the headlamp/horn stalk so that it is easy to sound the horn accidentally or to switch the headlamps to main beam at the very moment when the driver wishes to switch them off. On the Renault the stalk-operated minor controls are quite good, although the heater controls are less easy to find in a hurry but the instruments are poor, the speedometer being crude-looking and roughly calibrated while the mileometer is deeply recessed and difficult to see. Most of our drivers found the strip layout of the 1800's instruments effective and pleasing and the facia-mounted switches not difficult to reach (rocker switches are now used on the facia). American-styled deeply—some thought excessively—recessed instruments are fitted to the Victor, and the various knobs beside them were rather a long stretch for the driver, particularly when wearing seat-belts.

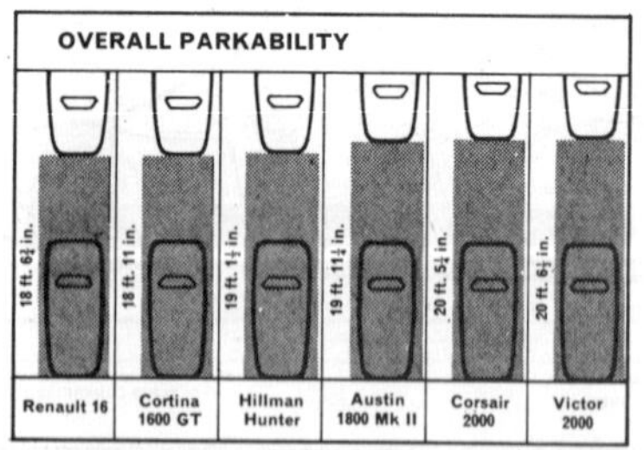

Heating and ventilation

All the cars had face-level fresh air vents on the facia and all but the 1800 extractor vents at the rear. Many of the face-level vents were badly located on the driver's side tending to direct cold air on to the right hand, the Victor and Hunter being slightly worse than the two Fords. On the Renault, however, the vents were better located and this car also had the most powerful and easily controlled heating system. The BMC 1800 had a rather feeble heater.

Noise

Road noise was not obtrusive in this group of cars, although it was low in the Renault and noticeable in both the Hunter and the 1800. The amount of wind noise, however, varied widely. It was low on the Renault and 1800, moderate in the Corsair and Hunter, but excessive and very obtrusive on the Victor and Cortina. There was less difference in engine noise, only the Corsair, with its rough, boomy engine note and the Cortina, with its fussiness at high revs being obtrusive—although the Victor was almost as noisy at high speeds—while both the Renault and 1800 were quiet.

'Motor' road tests of the cars in this group were published as below

Ford Cortina 1600E	February 24 1968
Ford Corsair 2000E	April 8 1967
Austin 1800 Mk. II	May 25 1968
Hillman Hunter	October 15 1966
Vauxhall Victor 2000	December 23 1967
Renault 16	February 5 1966

Back numbers or reprints are available from 'Motor', 40 Bowling Green Lane, London, E.C.1, price 3s 6d each post paid.

● I WAS quite sure I would prefer the Cortina GT out of this lot, but I discovered two facts of which I had previously been almost unaware: 1) that the Austin 1800 is an extremely good driver's car and, 2) that the Renault 16 now handles much better than I remember from earlier versions. The result is that my final choice comes to the Austin 1800 with the 16 a close second. Perhaps I prefer front wheel drive.

This 1800 really was a surprise when compared with the first of the line; it gets round corners astonishingly quickly and stably and it is almost impossible to get into trouble on the road—it can get into a lurchy neutral steer if you throw it into a corner on a trailing throttle, but with less desperate treatment it is about as fast and a good deal more comfortable. The understeer never worries me since I don't use full power all the way through a corner; the only unfortunate feature here is a fair amount of tyre squeal and apparently fairly rapid front tyre wear. The steering is not as heavy or as low geared as I had remembered, the performance is a lot smoother, quieter and faster and it is only the fact that it is unnecessarily large for me that doesn't put it well ahead of the rest.

It is the more compact size and feel of mechanical refinement of the 16 that makes an immediate impression. It is very quiet and well insulated and the roadholding is unbelievably good once you realize that the high roll angles don't mean that inversion is imminent; in fact this one on Michelin X tyres could be made to reach a final tail-sliding oversteer which was completely controllable. It was fun to drive and really quite brisk as well as being extremely comfortable on all sorts of road surfaces; the superb seats wouldn't look out of place in a Habitat sun lounge and they still manage to give a very comfortable driving position; the pedal arrangement made heel and toeing impossible.

The Cortina GT is still fun but it is neither as fast nor as good at roadholding as it might be, a fact which only becomes really noticeable in a direct and immediate comparison. Wind noise is about as bad as on the previous Cortinas, but the car is generally well equipped and about the right size for me. Curiously the Corsair 2000, which I have never

Testers' personal choice

Sequel to Group test No. 5 last week

This time, a draw. Victor and 1800 run neck and neck

liked as much as the better balanced in-line engined car, is almost a better GT although its final oversteer at higher cornering forces is less well behaved; I will never love its engine although it is powerful.

The Victor is a very fair driver's car, with a well-behaved live axle but ruined for me by low-geared steering. The rest of it is certainly a well balanced modern design which I could happily live with but, like the Corsair, it is unnecessarily large for me.

I think the Hunter is an extremely good family car but it doesn't begin to appeal to me; the design is "basic developed to modern" which makes it a thoroughly good but somewhat dull driver's car. There would be room for it in the family, but not as my sole choice.

Michael Bowler

● THE first thing you notice when you get into a strange vehicle, and one to which I am particularly sensitive, is the seat and driving position. A lot of people like to sit upright and close to the steering wheel but I'm not one of them. This rules out the two Fords even if they did have the best gearboxes and the best heel-and-toe pedal arrangements. The Corsair also had a very good ride and an extremely lively performance (from a rather harsh engine) but the hard ride and high wind noise of the Cortina GT made it unattractive to me.

That leaves four; just to be difficult I now propose to reject the one with the best driving position of the lot—the Hunter—because this example seemed under-damped at the rear and it suffered badly from wheel hop on bad surfaces. The Austin 1800, on the other hand, probably has the best roadholding of the lot and it has developed into a very desirable car—it is quick and it has a very smooth quiet engine except at tickover when the transmission idler gears rattle loudly. But it has more road noise than anything else, the gearchange is only "fair" and it is over-stabilised for my taste; if you drive it very fast understeer can develop into front end breakaway, particularly in wet weather.

Like the 1800, the Renault 16 also has outstanding roadholding (the two cars with i.r.s.); if it rolled less and had higher intermediate gears it would win my nomination. As it is, the Victor gets it because it has a very good ride, the best balanced handling in the group, little roll, a good though noisy gearbox, good driving position and adequate seats. It only achieves this distinction marginally because the suspension transmits road feel rather harshly and the engine is mechanically rather noisy at low speeds and again at speeds over 70 m.p.h. If I were doing a lot of cruising in the 80-90 m.p.h. region (as in France?) then I would be tempted to change my mind and opt for the quiet, refined, well-sprung Renault 16 especially as its handling was so much better than that of the last one I drove.

Charles Bulmer

● ONE of my pet ambitions is to persuade a computer operator to programme his transistors to calculate from suitable drawings the ratio of the total volume of a car to the volume inside it available to the passengers and luggage. The result would be a figure of merit indicating space utilization efficiency. I suspect that cars like the Ford Corsair and the Vauxhall Victor would not come out too well when judged in this way, that the more compact Hunter and Cortina would prove to be reasonably roomy with the Renault 16 even better but that the BMC 1800 would show a tremendous advantage.

This is just one of the reasons why I like the Austin 1800 the best and feel that this group test is a vindication—if vindication is needed—of its good qualities. The 1800 not only competes well in this group but would do well in the next group above it against the biggest models produced by the other major manufacturers, which tend to have a lot of metal without a proportionate amount of internal space.

One advantage of this compactness—coupled with an advanced suspension system—is that the 1800 handles like a real car rather than a boat with a broken rudder which is about how many very big cars feel for me. The 1800 has had to battle against a bad reputation caused by heavy, low-geared steering, badly located minor controls, ugliness and unreliability. Most of these faults have now been cured or mitigated, and as such it is outstanding value for money—if I had to buy a big car this would be the one, but perhaps with a Downton conversion.

Tony Curtis

● ON THE basis that a personal choice is usually the car that fulfils requirements most completely, my choice in previous groups has usually been a toss-up between two or three cars of fairly similar design. But in this group my two favourites could hardly be more different: the Victor 2000 and the Renault 16.

Ignoring its exceptionally good looks (because road testing should disregard aesthetics), I think I would just prefer the Victor for its greater performance, excellent driving position and visibility, good ride, and safe, neutral handling. It has a lot of appeal for me despite an engine which is noisy at high revs, particularly on the overrun, a vague notchy gearchange and far too much wind noise. And while I admire Vauxhall's efforts to lighten controls, I would gladly accept slightly heavier steering in return for higher gearing and more feel.

The Renault, benefiting greatly from refinements in the latest models, must be one of the most relaxing cars in its class, with the occupants isolated from road, wind and engine noise to an uncanny degree. The steering is a little lighter than I remembered, the ride is excellent and it is one of the few fwd cars with which genuine oversteer can be induced, though with more lurch and roll than I like. We had little cause to exploit the seating permutations of this versatile saloon/estate but if we'd had to carry the sheep instead of just avoiding them, this would probably have been my best buy.

Oversteer is still a term unknown to my third choice, the Austin 1800, which will storm round any corner in its path in a greater or lesser degree of understeer determined only by speed and radius; none of the sudden transition when you lift off, as characterized by some of its smaller stablemates. Recent modifications seem to have both lightened and quickened the steering. I thought that the ride seemed a little different too; it is both firmer and more noisy at low speeds and under certain conditions an uncomfortably sharp oscillating motion is set up. I have not noticed this on 1800s before and it might be partly attributable to the revised seats.

Of the two Fords I preferred the Corsair despite the noise and roughness of the V-4—another which appears to have escaped Harley Copp's purge on NVH (noise vibration and harshness). It has a lot more prod than the in-line unit, particularly low down, a slightly better ride and rather more entertaining handling with a delightfully progressive oversteer. The Cortina suffered from appalling wind noise and was fussy at speed but has more comfortable seats with a just tolerable degree of rearward adjustment (unlike the Corsair).

I would like to think that our Hunter was not a good example—a back axle pawing the air for much of the time is not only uncomfortable but seriously detracts from cornering power on all but the smoothest surfaces. I like its driving position and seats, the only one of the six with rake adjustment, the gearbox is equally as good as the Fords', and with overdrive cruising was almost as restful as with the Renault.

Jim Tosen

Giant comparison by WHEELS staff

FOUR FURIOUS FOURS

4

Fiat 125—Mazda 1500 SS—Peugeot 404—Cortina GT1600

This time we rate four of the most desirable of the considerable machinery selling in the six-cylinder compact area — and selling only on quality and performance. These are all drivers' cars — so fasten your belts!

Driven hard at the same speed around a marked corner at Hardie-Ferodo, all four show remarkably little roll, excellent stability.

THE comparison of four under-$3000 high performance cars was largely prompted by the release of the Fiat 125. The question was immediately posed — what other cars can offer the performance, comfort and handling that will satisfy the more mature enthusiast in terms of four-door transport plus handling and performance for occasional club competition or a quick country-interstate trip?

The Cortina GT 1600 is the most obvious rival but we had to take a quick peek at prices to find that the Toyota Corona 1600S, Péugeot 404 and Mazda 1500SS were all sporty sedans in a similar price category. As the Corona 1600S is available in only small quantity and at the time of the comparison no test car was around the decision was made easy. The Ford, Peugeot and Mazda are all cheaper than the Fiat, which with the Mazda is fully imported and hence suffers heavy tariffs and consequent overpricing. The 125 should be in the $2500 bracket in the same way its brother the 124 should be under $2000 rather than over. The Mazda runs to $2620, the Peugeot 404 $2550 and the GT $2530, which puts a blanket of $368 over the four, a wide difference for a WHEELS comparison but we feel that buyers of enthusiast machinery are normally prepared to pay the extra for the car they want.

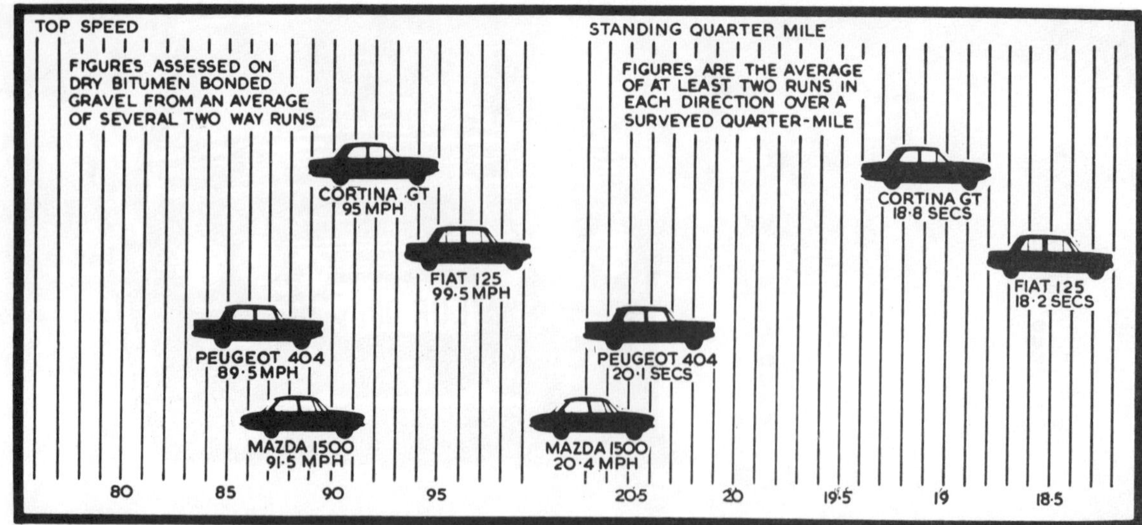

At first, the comparison may seem unfair to the Peugeot and Mazda, but scrutiny of the specifications shows how alike the four are. All but the Mazda are 1600s, all have horsepower ratings between 80-90 with weights from 18 to 21 cwt. The Fiat has only recently been released in Australia, the Cortina GT now uses the 1600 crossflow engine with the Mk 2 body introduced late 1967, the Mazda 1500SS was released early in 1968 having more interior goodies and slightly more power than the normal 1500 Deluxe and the oldest of the four which is basically the same as the car first released in 1962, the Peugeot 404 received its latest updating late in 1967 with new facia layout and different gearbox ratios.

Driving positions: 1. Fiat driver sits well behind canted wheel, gearlever a little far back. 2. Mazda relationship is excellent, but gearlever knob is too close to seats. 3. Peugeot has least conventional position. 4. Cortina wheel is too close.

GENERAL

As all cars had been recently tested we used performance figures from the respective road tests but took specific dimensions and equipment details when the four were

1. Fiat leaves large area uncleaned bottom right — which we rate as dangerous. 2. Mazda cleans big area — but oh, that centre vee! 3. Peugeot has shortest blades of all, leaves at least three blind spots. 4. Cortina is the best of a bad lot.

collected for photography. All cars supplied were in amazingly standard condition — as road test cars go — with only the Cortina sporting a non-standard radio and three point belts where lap belts are standard. Neither the Fiat or Mazda have belts but the Peugeot scores with three-pointers. The cars were supplied by Fiat (Australia), Renault (Australia) who distribute Peugeots and the Cortina from Ford (Australia). The Mazda was supplied by the enterprising North Sydney dealer Eurocars.

Waiting lists for the newest of the four, the Fiat 125, bear testimony to its astonishing value for money despite a high price tag. When arranging the Cortina GT for photography we found that few dealers had cars readily available as most were on order. The demand for the 1500SS also exceeds supply but the Peugeot which is locally assembled at the Heidleberg, Victoria factory of Renault is finally immediately available.

PERFORMANCE	Cortina GT	Fiat 125	Peugeot 404	Mazda 1500SS
Top speed	95.0 mph	99.5 mph	89.5 mph	91.5 mph
Standing ¼ mile	18.8 secs	18.2 secs	20.1 secs	20.4 secs
0-50 mph	9.5 secs	8.0 secs	12.9 secs	12.0 secs
30-50 mph	10.2 secs	9.7 secs	9.7 secs	12.4 secs
Fuel consumption	26.2 mpg	22.1 mpg	28.9 mpg	27.0 mpg
Max in gears:				
First	35 mph	35 mph	28 mph	27 mph
Second	50 mph	55 mph	48 mph	45 mph
Third	70 mph	80 mph	73 mph	72 mph
Top	95 mph	99 mph	89 mph	91 mph
CALCULATED DATA				
Final drive ratio	3.9 to 1	4.1 to 1	4.23 to 1	4.111 to 1
Mph per 1000 rpm	17.1	16.2	17.7	17.5
Piston speed at max bhp	2580 ft/min	3000 ft/min	2671 ft/min	2758 ft/min
Lbs (kerb) per (gross) bhp	22.2	25.0	28.0	27.5

1. Fiat engine bay is cluttered, disorganised. 2. Mazda also has a lot of plumbing but is relatively orderly. 3. Canted Peugeot engine gives good working room. 4. Cortina engine is simplest, neatest of all.

ENGINES

The Fiat bears up as a poor man's Alfa sporting a twin overhead cam engine which is seconded by the Mazda with a single overhead cam. Strangely both shape up as being exactly square and both run in five bearing crankshafts. Both are free-spinning engines and develop their maximum torque high up in the rpm range which means no laziness with the gearbox. However the Fiat's engine is not cammy and can accelerate smoothly from reasonably slow speeds to give deceptively quick passing acceleration. The Mazda's engine in comparison (86 bhp to the Fiat's 90) is yet another case of Japanese ponies. There is no reason to suggest that either overhead cam engines are likely to be fussy in service. Both have been developed for reliability and run exceptionally smoothly when idling with the toothed belt drive Fiat engine making slightly more clatter than the chain drive Mazda. The Peugeot and Cortina must be grouped for their pushrod engines as the Mazda and Fiat are for their ohc units. Both are oversquare and have lower piston speeds than the ohc engines. Their respective maximum torque comes in at lower rpm with the Peugeot having greatest torque of the four at the lowest speed. This explains its exceptional 30-50 mph passing acceleration time of 9.7 secs which is on par with the more powerful Fiat. As we write this, news arrives of yet another Peugeot victory in the East African Safari rally. The Peugeot record for engine reliability and solidarity is legendary and must lead the four on this point. It is perhaps the noisiest of the four when worked hard. The Cortina uses the well proven five bearing engine but now capped with a cross-flow cylinder head for an increase of 5 bhp over the old GT. The Mazda is alone in using twin carburettors where the other three have sought twin choke single instruments as the best compromise for performance and economy. It is a tribute to the Mazda's head design that despite the twin carbies which are twin choke units the fuel consumption is second best. Accessibility to the engine compartment is better on the two pushrod engines than the ohc cars for that reason.

TRANSMISSION

All four use four speeders with full syncromesh. The Fiat and Cortina are hard to separate.

Continued on page 52

SPECIFICATIONS	Cortina GT	Fiat 125	Peugeot 404	Mazda 1500SS
Cubic capacity	1599 cc	1608 cc	1618 cc	1490 cc
Bore and stroke	80.9 x 70.62	80 x 80	84 x 73	78 x 78
Bhp at rpm	88 at 5400	90 at 5700	80 at 5600	86 at 5500
Torque (lb/ft) at rpm	96 at 3600	94 at 4000	97 at 2500	86.7 at 3000
Compression ratio	9.0 to 1	8.1 to 1	8.3 to 1	9.0 to 1
Carburettor	Weber	Weber or Solex	Solex	two Stromberg
Transmission	four, all syn	four, all syn	four, all syn	four, all syn
Gear lever location	floor	floor	column	floor
Suspension front	coils, strut	coils	coils, strut	coils
Suspension rear	leaf	leaf	coils	leaf
Tyre size	165 x 13	175 x 13	165x380 (5.90x15)	6.45 x 14
Steering type	recirc ball	worm and roller	rack and pinion	ball nut
Turns l to l	4.5	four	3.7	4.2
Turning circle	30 ft	35 ft	30 ft	32 ft
Brakes (swept area)	na	na	168 sq in.	na
Brakes front/rear	disc/drum	disc/disc	drum/drum	disc/drum
Price	$2530	$2898	$2550	$2620
OVERALL DIMENSIONS				
Wheelbase	8 ft 2 in.	8 ft 3 in.	8 ft 8¼ in.	8 ft 3 in.
Track front	4 ft 4.5 in.	4 ft 4 in.	4 ft 5¼ in.	4 ft 4 in.
Track rear	4 ft 3 in.	4 ft 3 in.	4 ft 2¾ in.	4 ft 4 in.
Length	14 ft	13 ft 10 in.	14 ft 7 in.	14 ft 4 in.
Width	5 ft 4.75 in.	5 ft 3 in.	5 ft 3⅛ in.	5 ft 4 in.
Height	4 ft 8.5 in.	4 ft 7 in.	4 ft 2½ in.	4 ft 7 in.
Ground clearance	6.6 in.	6.9 in.	6 in.	7 in.
Kerb weight	17.75 cwt	19.9 cwt	20.5 cwt	21 cwt
Fuel tank capacity	10 gals	10 gals	12 gals	11 gals
SPECIFIC DIMENSIONS				
Boot lip height	2 ft 5½ in.	2 ft 9 in.	2 ft 7 in.	2 ft 4 in.
Depth of boot	1 ft 7 in.	1 ft 6 in.	1 ft 6¼ in.	1 ft 6 in.
Mean interior width	4 ft 2 in.	4 ft 1 in.	na	3 ft 11¾ in.
Driver's window width	1 ft 6 in.	1 ft 7½ in.	2 ft 1 in.	1 ft 6½ in.
Front door aperture	2 ft 9 in.	3 ft	2 ft 10½ in.	2 ft 10 in.
Legroom rear	na	7¼ in.	10 in.	4½ in.
Front seat travel	3½ in.	7 in.	7½ in.	6½ in.
Steering wheel diam.	15 in.	15½ in.	16½ in.	16 in.
Chest to wheel boss centre	1 ft 6 in.	1 ft 7½ in.	1 ft 8 in.	1 ft 8 in.
Rear vision mirror width	5¾ in.	8½ in.	7½ in.	8 in.
Effective glovebox width	1 ft 0 in.	11 in.	1 ft 2 in.	1 ft 2¾ in.
Effective glovebox depth	8½ in.	3 in.	8 in.	5½ in.
EQUIPMENT				
Heater/demister	yes	yes	yes	yes
Windscreen wipers	one-speed	two-speed	one-speed	two-speed
Windscreen washers	manual	power	manual	power
Reversing light	yes	yes	no	yes
Ashtrays	1f, 2r	1f, 2r	1f, 2r	1f, 1r
Parcels shelf	yes	yes	no	yes
Grab handles	no	1f, 2r	no	2r
Childproof rear door lock	yes	no	yes	no
Headlight flasher	yes	yes	yes	no
Seat belts	lap	no	lap/sash	no
Window winder turns	2¾	5¾	5½	2¾
Instruments	sp, 10th odo, fuel, temp, oil, amps	sp, odo, 10th trip, fuel, temp, clock	sp, odo, 10th trip, clock, amp, fuel, temp	sp, 10th odo, temp, fuel, clock, amps
Tachometer	yes	no	no	yes
Choke system	man	man	man	man
Alternator	no	yes	no	yes
Bonnet stay	manual	manual	counter-balance	manual
Through flow ventilation	yes	yes	no, facia vents only	no
Opening ¼ vents	no	yes	no ¼ vent	yes
Lockable glovebox	no	yes	yes	no
Armrests	2f, 2r	2f, 2r	2f, 3r	2f, 3r
Cigarette lighter	no	yes	yes	yes
Radial ply tyres	optional	standard	optional	no
Repeater side indicators	no	yes	no	yes
Radio	no	no	no	yes

FOUR FURIOUS FOURS

Continued from page 50

The Fiat's is typically Fiat with a slight feeling of notchiness and a definite "gate" to work through. The Cortina's gears can be swapped at will as fast or as lazily or as slap-dash as possible with or without the clutch — well, just about — and there doesn't even look like being a protest. It must rate as one of the best gearboxes for novices it is so forgiving — in the hands of an enthusiast . . . ! The clutch-gearbox integration is excellent on the Fiat but prefers enthusiastic driving as the clutch take-up is close to the floor. Lazy driving will often find the clutch engaged without the next gear selected. The Peugeot up to the current model has always been labelled as one of those funny French cars if for nothing else but the strange gear shift pattern. For the latest model they have reverted to convention and used a simple H-gate similar to column shift Hillmans, Austins and Japanese cars but just to confuse the issue have reversed the direction. That puts the first-second gate nearest the facia. Whatever, it is an improvement over the old system especially for simple minded road testers that climb from one car into the next and expect to find the gears where they should be. The actual change is excellent with a slight French mushiness, and Continental "gate" feel similar to the Fiat but fast and easy to find. The pattern has been aligned so swapping across the gate 2-3 or 3-2 is virtually a straight movement and not "up-across-up". The Mazda falls a sad last in the gearbox department. We have no quarrel with the actual box as there is no reason to suspect it is not every bit as reliable as the other three but the change mechanism is notchy, slow, has heavy syncros and a change lever which is definitely not FETH (a simple term meaning falls-easily-to-hand). First requires a good stretch while a snatch change back to two will land you back into the front seat cushion or into the knee of the front passenger — a dubious situation depending on sex of said passenger. But if it is slow and heavy, the syncros cannot be beaten which suggests that the unit is robust.

PERFORMANCE

For those buying performance the undisputed leader of the pack is the Fiat. Not only will it go zot into the middle distance but it does so in a seemingly unfussed way to the passengers. Its acceleration is quite deceptive with engine noise only disturbing the interior in the higher rpm ranges where it tends to blend with general wind-road noise and go unnoticed. Good high end torque gives the Fiat long open road legs and only the steepest hills require three which must be used early to maintain rpm and torque. The 125 presents a very good argument for just dashing interstate for the "helluvit" or a very comfortable argument if long tripping is a necessity. As a long distance tourer the Peugeot rates second and not far behind the Fiat. In their respective countries of origin the Fiat is really a class ahead of the Peugeot. Thus the Peugeot appears to work harder and be noisier than the Fiat when touring. But the engine is a willing hard worker and can be stirred along at maximum speeds unconcernedly. With better breathing the 404 would go considerably harder which is a fair reason behind the success of the Kugelfischer fuel injected Peugeots. Its natural touring gait is 75-80 mph where it will sit all day and maintain high point-to-point averages. In town its low down torque gives excellent top gear pulling. In terms of straight line performance the Cortina GT sits between the Fiat and the Peugeot but on the open road fast driving seems harder work than in the Peugeot. Seats and general comfort could be the factors. In town full use needs to be made of the gearbox which really presents no bugbear especially to Cortina GT clientele. The Mazda is a bit left behind when it comes to the Great Point To Point Race (don't get us wrong, we're not suggesting a four abreast flog from Sydney to Melbourne — merely a figment of the imagination for evaluating relative performances). As said in our original road test of the SS maybe we expect too much from the car as a sports version of the well liked 1500 Deluxe. The SS rates more as an ME (more equipment) with extra instruments, comfort, disc brakes and more horsepower to offset the extra weight. As such, try as we did, the performance figures were little better than the normal 1500. Unfortunately the car has an extraordinarily "dead" feel in terms of go, brakes and handling, a feature that will not endear it to the enthusiast. However it does have equipment and comfort to place it in the sports sedan bracket. On fuel consumption there is little to pick between the cars and a slightly worse figure for the Fiat could be largely attributed to hard driving. Not that we excuse it, for even the sanest owner will be tempted to use the 125 hard.

BRAKES

The Fiat wins here, pretty well nose down — if you'll pardon the phrase. Servo boosted disc brakes all round grab the 125 like a giant hand. The pedal pressure is light and reassuring with a fair travel from the point of original application to full on. This gives great control over braking power. The OE Pirelli Sempione tyres complement the discs for ridiculously small stopping distances. With discs up front, power assisted, and rear drums the Cortina rates next. Its pedal pressure is also firm and reassuring for good stopping power without giving the over-confidence that the giant hand of Fiat and Renault four wheel disc set-ups encourage. The all drum system of the Peugeot must be one of the last outposts for those who steadfastly cry good drums are more reliable than temperamental discs. And that cry is not unfounded for the 404 has excellent brakes. The pedal requires a heavier tread than the discs of the 125 or GT but stop the car in a straight line, are free from fade except in ultra extreme conditions and we would bet they'd be the most easily maintained of all. The SS has a power assisted disc front-drum rear system which is well up to the standard befitting a sports sedan. The pressures are heavier than the Cortina's disc/drum set-up but the stopping power is more than adequate for the car's performance.

HANDLING, RIDE

As all four are quite conventional in layout — front engine, rear wheel drive — their individual handling traits are mainly decided by power and the rear wheel location. The Fiat with greatest torque available at the rear and put to ground through a rear axle located on soft two leaf springs, hefty upper torque rods and strong rate dampers is very controllable. Only the slightest trace of front end plough or understeer is evident on the highway and on loose surfaces this is easily transformed into controlled opposite lock slides depending on the power applied. It is very much an enthusiast's type of handling. Ride is well damped and competes with the Pug as best of the four. In interests of handling the ride is a bit taut and bumpy over indifferent surfaces. The Peugeot has also been set-up on a well located coil sprung rear end so that with rear pressures 2 lbs lower than front the car can be made to slightly oversteer even on bitumen for very fast open road cornering. The suspension is quite soft and despite its excellent location there is often violent patter set-up by the inside rear wheel when cornering on rough surfaces. However the soft suspension gives a comfortable ride with mild body lean but excellent for tireless tripping. The soft sus-

CONTINUED ON PAGE 96

24,000 mile staff car report

Ford Cortina GT

by Philip Turner

'The car has been thoroughly reliable and still retains its performance'

BECAUSE I WAS getting through 12,000 mile tests at such a rate that I seemed for ever to be running-in, it was decided that I should keep my test cars for the full 24,000 miles, rather than hand them over to a colleague to complete the second 12,000-mile stint. So it inevitably followed that once this had been agreed for the Ford Cortina GT, I found myself covering fewer miles than usual in my staff car and many, many more in cars lent me by friendly manufacturers! As a result, the Cortina of which I took delivery on March 28, 1967, is still with me some 18 months and 28,000 miles later. It is the Mk.2 version but powered by the old 1,500 c.c. engine and not by the current 1,600 c.c. bowl-in-head unit. It was sufficiently late in production to have the revised gearbox with the higher second gear and was delivered to me equipped, as requested, with Pirelli Cinturato tyres on the wider $4\frac{1}{2}$ rims which transform the road holding of all Cortinas, especially in the wet.

November, 1967, and the 12,000-mile mark appeared more or less together so that the **12,000–13,000 miles** period was spent in darting about in Britain, rather than on long, Continental runs. Not that, as some suppose, sports editors hibernate throughout the winter months. **13,000–14,000 miles** (December/January) were also amassed on British roads. The car was dead reliable, we poured in petrol and oil as and when required and just went motoring steadily on with nothing of the slightest significance to report.

14,000–15,000 miles (January/February). Not much to report at this stage either. Thick snow early in January once again showed that the Cortina on the wider, $4\frac{1}{2}$in. wide wheels has excellent traction—once I had dug the car out of its garage. Otherwise, there was just the normal running about, including a two-day run up to the Midlands. The car spent some time waiting for me at airports; at Gatwick while I went to see the new Matra V-12 GP engine and at Heathrow during the fortnight I was away reporting the Monte-Carlo Rally and visiting Ferrari.

15,000–16,000 miles (February/March). No maintenance required, we just motored. To Brands to see the then new Formula 1 McLaren on test, to Thruxton on a bleak, day of icy wind to take my first look at the new circuit the BARC were just completing, to Brands again for a special Ford demonstration of the competition versions of the Escort.

16,000–17,000 miles (March). Performance became rather sluggish, starting a little less certain than usual and the tappets were clickety clacking—all signs that a 5,000 mile service was due. So at 16,672 miles the car went to Highbury Corner Motors. The rev-counter—a replacement of the original—had gone on the blink. Sometimes it did, and sometimes it didn't. A thump with the closed fist on its vacant face would on occasion send the needle swinging round the dial as it came to life again, but after a time it would lose interest and the needle would go into a decline so that, to a casual passenger, mine was a most unusual Cortina with a low speed engine that turned at only 1,800 r.p.m. at 70 m.p.h. I gather mine was not the only Cortina GT suffering from this problem,

Continued on the next page

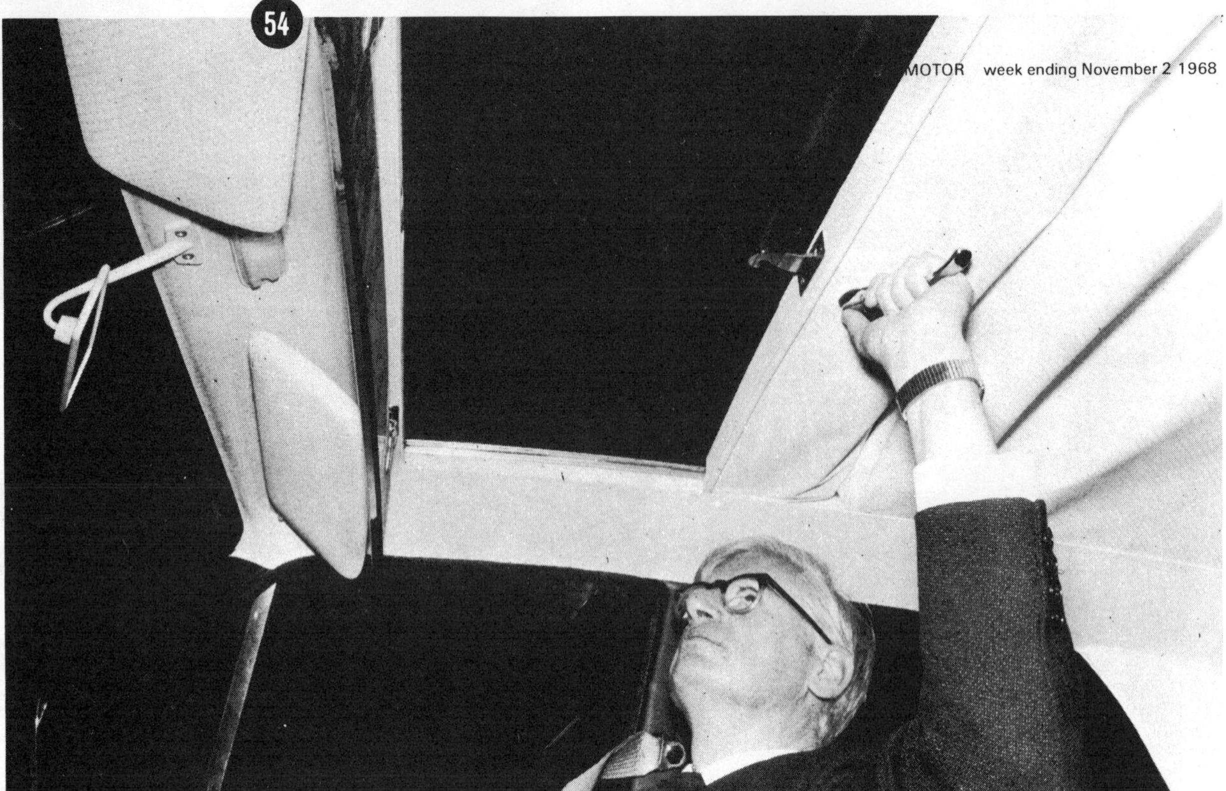

Opening the Golde sun roof is a swift, single-handed operation. For town use, it can be locked in the half-open position shown.

The oil filler cap has given rise to some pretty strong curses in a wide variety of tongues for it is difficult to remove. Fortunately, oil consumption seems to average about a pint per 500 miles, even when driven hard.

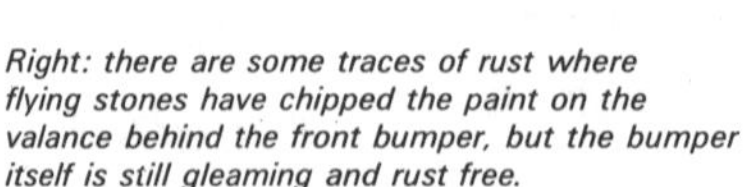

Right: there are some traces of rust where flying stones have chipped the paint on the valance behind the front bumper, but the bumper itself is still gleaming and rust free.

24,000 mile report

continued

for replacement rev-counters never seemed to be in stock and, frankly it did not seem worth the trouble of taking the car in specially for a new instrument to be fitted when there was every likelihood that the replacement would prove just as erratic before long.

When I collected the Cortina on the evening of the day I had left it to be serviced, the performance had regained its usual sparkle. I sometimes wonder if the present servicing routine at 5,000 mile intervals is not an unmixed blessing. Not many cars that I have owned have run for 5,000 miles without a gradual tailing off in performance. A plugs, points and tappets service would usually restore the engine to tune.

It was in March that motor racing came back to life again in Britain with the Race of the Champions at Brands Hatch, so from now on the monthly mileage began to climb again, and once more the windscreen was decorated with the tatty remains of paddock car passes which refused to unstick completely when torn off. **17,000–18,000 miles** (April). Brands Hatch (BOAC 500), Snetterton (Guards 500) and Thruxton's opening meeting packed this thousand miles into the first two weeks of April. **18,000–19,000 miles** (April/May). Two major modifications were carried out during this period. On May 1 at 18,740 miles the car went to the Allard Motor Co., Ltd. workshops in East Putney for a Golde Continental sun roof to be fitted. When we first discussed having this major operation carried out I was not all that enthusiastic, for I feared that cutting a large hole in the roof might weaken the body shell which would then quiver like a jelly when driven hard over Continental roads and ruin the road holding. However, further investigation showed that transverse steel bars are welded to the body sub-frame to ensure that the body remains as rigid as ever, and that the Ford Motor Company had given their blessing to this particular roof. How the operation is carried out is dealt with on page 51. From the on-the-road point of view I have never since ceased to rejoice in this new hole overhead from the moment I collected the car from Allards on that May afternoon and slid the roof open.

I do believe that the last car with a sun roof I had driven was the family Ford 14.9 of 1934 vintage which, in spite of the fact that it was named by the apparently humourless Dagenham characters of those days the BF model, served us well for several years and on one occasion found itself lapping the Nürburgring with a much, much younger Turner at the wheel. Its sun roof was not its strong point, however, being much inclined to leak in heavy rain and to become firmly stuck in the closed position so that brute strength and even more brutal curses were needed to open it.

By contrast, the Golde roof requires but a featherlight touch to open or close it, and the whole operation can be carried out single handed by the driver during a traffic-light stop. Nor has it ever leaked, not even during some of the torrential downpours of this disastrous summer. I had some reservations on whether or not it would stand up to being motored on Continental motorways at high speed in the open position, for should it suddenly tear away and fly up like a sail it could well throw a car completely out of control. I was assured by Allards that there was not the slightest risk of this happening so long as the roof was opened to its full extent and not

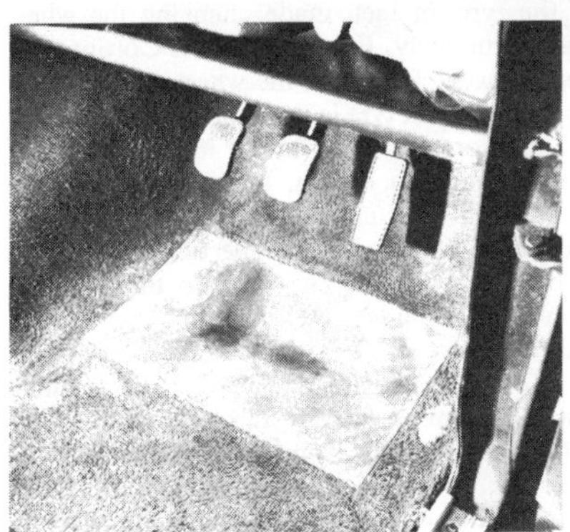

Turner passes by. This photograph shows that the Cortina still retains both its performance and its good finish despite 18 months' hard motoring.

The carpet in the front cockpit is beginning to wear somewhat thin in places, but the appeal of the Cortina GT lies essentially in its excellent performance rather than in its luxury trimmings.

Right: in spite of motoring over salt treated, snow covered roads last winter, all four hub caps are still rust free and gleaming.

to the half open position designed for town use only. I have since driven for mile after mile on German autobahnen with the roof open and the speedometer well over the 90 m.p.h. mark without any untoward happenings, and it still operates as freely as ever and is still completely leak proof.

Another reason for retaining this Ford for as long as possible is that the very thought of returning to a non-opening tin roof fills both my family and me with the utmost despondency. It is surprising how often we run with the roof open, not only on long runs but also on the occasional home-to-office run. And how one enjoys the open roof when collecting the car of an evening from the office car park, or the paddock, in which it has stood for hours in a hot sun, then the ability to open the overhead hatch and let in cool, refreshing air is a boon and a blessing.

The following week the car went into Clarke and Simpson, Ltd. of Sloane Square for another major modification, the replacing of its front suspension legs with the Cortina-Lotus type. As I remarked in the 12,000 mile report on the Cortina published in our February 10, 1968, issue, I was not too happy with the car at speed on any but the smoother French roads. The front especially seemed far too agile and one wondered in which direction it would dart next. I had heard that Cortina-Lotus front struts with their different damper settings improved matters enormously, but I had experienced great difficulty in finding a Ford dealer to carry out the conversion for me. Either they did not stock Lotus type front struts or they persisted in assuring me that there was no difference between the standard struts and the Lotus ones. Then by chance I came across a Clarke and Simpson advertisement in the Sutton and Cheam Motor Club magazine. Their David Sutton who is in charge of the competition department where they prepare Cortinas for rallying at once knew what I was talking about and put the whole operation in hand for me, with the result that I drove the Cortina into their Sloane Square workshops on Monday morning and collected it again that evening with the work completed.

At first, I noticed no immediate improvement, but as I neared home it dawned on me that I was driving the car in a much more enterprising fashion than hitherto, just as though it were a Mini, in fact. Eagerly I awaited my first foreign expedition which should have been to Le Mans but in fact was to Zandvoort owing to the postponement of the 24 hour race until September. Since that first Dutch trip, I have covered many miles in Belgium, Germany and France, and to my mind there is no doubt that this change in front suspension struts has increased the car's cruising speed from a dicey 70 m.p.h. to a safe feeling 80 m.p.h. on the less satisfactory Continental roads. Nor is this only my view, for everyone else who has driven the car abroad has commented on its greatly improved handling. The cost of this modification was £15 and I must say I think it is cheap at the price and very well worth the having for any Cortina owner who covers a lot of miles on the Continent.

19,000–20,000 miles (May/June). Although the racing season was now very much in full swing, the mileage increased rather slowly during this period as the car sat in its home garage for a fortnight while I was in Monte Carlo and Italy. Then in the third week in June we set off together for Holland and the Dutch GP.

20,000–21,000 miles (June/July). The Dutch expedition ended in a slight disaster when at Rotterdam Airport the loaders drove the Cortina out on to the apron ready for insertion into the Carvair's maw—and then managed to slam lock it with the key in the ignition

24,000 mile report

continued

switch. Quite how they managed this, I never found out. Of more immediate interest was the fact that the spare key was in the locker under the central arm rest. We prowled round and round that motor car, the loaders, the British Air Ferries staff and the captain and second pilot of the Carvair, but no way could we find of bugGlariously entering it. The boot, of course, slam locks, there are no opening quarter lights, and tentative attempts at threading a hooked wire under the opening roof explained just why it is so leak proof—it is a very tight fit indeed once closed. In the end, we were forced to break the front quarter light on the passenger's door, but I should be interested to learn from a professional car thief of any alternative method of effecting an entry.

Soon after its somewhat over-ventilated return to England, the Cortina went to Highbury Corner Motors for its 20,000-mile service, which included new sparking plugs and new pads for the front discs. Persistent squeaks from the left front wheel on the way home were investigated by our Temple Press garage on arrival at the office the following morning and were found to be due to the new pads being fitted with so little clearance the front wheels could barely be turned by hand. This is not up to the usual standard of servicing provided by the company concerned, but it was their only major boob while the car has been serviced by them ever since they supplied it in the first place, and on the whole they have done a very good job. Certainly, it has never failed to return from one of their services with its performance greatly restored and sharpened.

21,000–22,000 miles (July). This thousand began with out and home runs to Brands Hatch for the British GP and ended with the car on a comprehensive swan around Germany which included a visit to the Freiburg hillclimb, then back to Stuttgart to spend a day with Porsche and the following morning in the Mercedes museum, after which we drove on to the Nürburgring for the German GP. Careful friends of mine load down their cars with spare parts before setting off on their one annual visit to the Continent. I'm afraid the only extra attention my cars receive on the eve of an extensive Continental sortie is a swift check on the tyre pressures and a peek at the dipstick to make sure there's plenty of oil, for adequate supplies of air and oil are, I believe, rather essential if a considerable motorway mileage is in prospect.

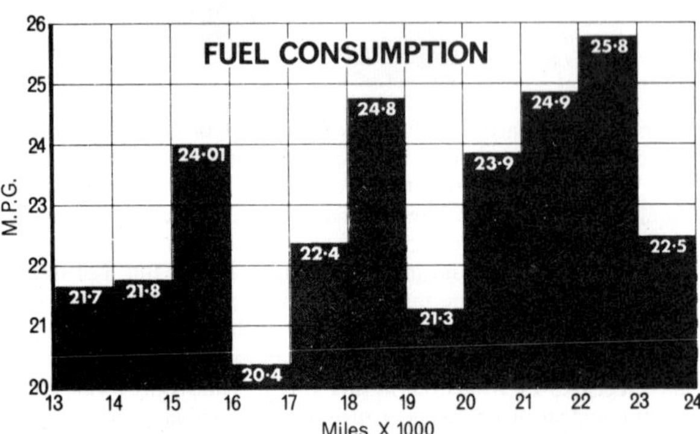

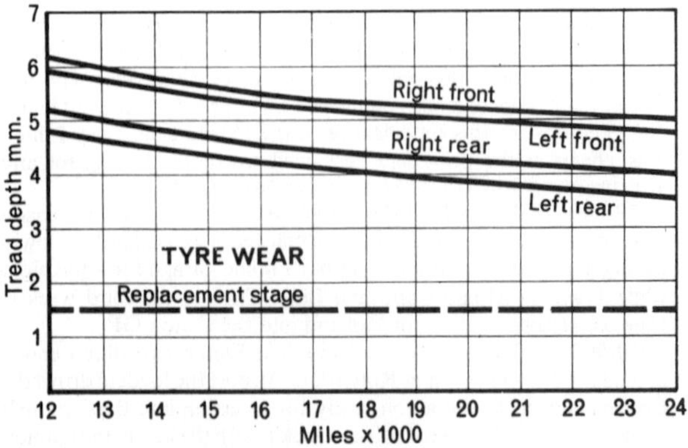

The Cortina fully justified my faith in its good works by giving not the slightest trouble during this 1,615-mile trip in 14 days. At the conclusion of a day spent at Porsche, Baron Huschke von Hanstein, Porsche's competition manager and chief of publicity, lent Maurice Rowe and me a Porsche each to drive, then led us to his house on the other side of Stuttgart at the wheel of my Cortina. This was a most amusing run—except when a massive dump truck began backing across our bows. I followed Huschke through the rapidly narrowing gap, but Maurice wisely clapped on the brakes before both he and the Porsche were cut off in their prime. Von Hanstein remarked later when seated on his terrace on how well the Cortina went and handled.

22,000—23,000 (July). This 1,000 miles swept by in Germany and the 23,000 mark was reached en route to the Nürburgring.

23,000—24,000 miles (August). Several laps of the ring were completed during the days before the race, none of them at all rapid as the idea was to seek good photographic positions rather than new lap records. In between circulating, the Cortina traversed some appalling tracks leading to various corners, mostly steeply uphill and carpeted with sharp stones. These may well have been responsible for a slight "moment" when homeward bound to Rotterdam airport. We were cruising at around 90 indicated on a Dutch motorway just before Arnhem when with a series of most disturbing sounds the left side rear tyre suddenly deflated. Maurice, who was driving at the time, slowed the car to a halt on the grass verge without any undue drama and successfully coped with any tendency to swerve across to the left, but by the time we stopped the tyre was a complete ruin enveloped in blue and very smelly smoke. The heat of the tyre, in fact, made changing the wheel somewhat difficult but, fortunately, Ford provide the Cortina with a proper and substantial jack. However, the wheel change completed, we stowed the still very hot tyre in the boot and continued to Rotterdam, Southend and home, without further incident.

The weekend after returning to Britain, I transferred the new replacement for the wrecked tyre and the unused spare we had fitted to the rear wheel after the puncture to the two front wheels and switched the two front wheels to the rear, so that my tyre graph really should finish at around 23,600 miles.

Apart from a wash, this was the only attention the Cortina received before setting off for Aberdeen to take ship for the Shetland Isles where we spent our summer holiday this year. In fact, the **24,000 miles** was registered by the mileometer just after crossing the "frontier" at Carter Bar. We stopped at Dundee on the outward journey, but drove straight through when returning home, covering the 556 miles between 9.10 a.m. and 9.15 p.m. I have no doubt many readers have done this journey in less time, but traffic congestion was very heavy on this fine Sunday, the car was driven fairly gently round the swervy bits in deference to my passengers and the very full load the boot had swallowed and we had stopped for coffee, lunch and tea en route.

The Cortina has now covered just over 28,000 miles and is still its reliable, rugged self. For a time the clutch would suddenly become rather fierce so that it was difficult to engage without judder, but after two or three gear changes it would resume its proper behaviour I intended having it seen to when I returned from Germany, but all of a sudden it ceased to misbehave in this manner and has been functioning perfectly ever since. Then a burst of acceleration in the gears would lead to a most startling sound at times due to a loose baffle or two in the silencer. Returning from Le Mans, a change in the engine note accompanied by occasional spitting back on the over-run showed that the exhaust gasket, too, was about to depart this life, so, on our return to London, the exhaust gasket and silencer were both replaced, and all is now well again.

The silencer and the front suspension struts are the only major items to be replaced in 28,000 miles. Otherwise, the car has been totally reliable and it still retains its performance. The Cortina GT is certainly not the quietest car, or the smoothest, most refined and most comfortable, but for anyone wanting a car that will retain its performance month in, month out, with the minimum of attention, and, moreover, one that handles a great deal better than many more expensive sports saloons, then the Cortina GT is indeed a Best Buy.

M

Autocar
Road test
NUMBER 2194

FORD CORTINA GT ESTATE 1,598 c.c.

AT A GLANCE: Special GT version of Super estate car. Performance above the 1600 class with brakes and handling to match. Close-ratio gearbox with unbeatable synchromesh. Good reclining seats (extra) and lots of instruments. Noise level reasonable. Fair ride in the front, but skimped back seat not comfortable. Vast load capacity. Expensive for a Cortina, but not for what it offers.

MANUFACTURER
Ford Motor Co. Ltd., Warley, Near Brentwood, Essex.

EXTRAS (inc. PT)

Reclining seats	**£25 11s 2d**
Metallic paint	**£6 7s 10d**
Radio	**£20 15s 4d**

PERFORMANCE SUMMARY

Mean maximum speed	96 mph
Standing start ¼-mile	19.3 sec
0-60 mph	14.2 sec
30-70 mph (through gears)	12.5 sec
Typical fuel consumption	26 mpg
Miles per tankful	210

IT is just over a year since we last tested a Cortina in estate car form. At that time, the two basic engine options were the 1,498 c.c. five-main-bearing unit inherited from the Mark I Cortina and an interim "destroked" version of the same engine, with a capacity of 1,297 c.c. Both engines, of course, were also used in the saloon models.

Last autumn, the new "Kent" series engines, with their cross-flow cylinder heads and bowl-in-piston combustion chambers, were introduced into the Cortina range. The smaller unit's capacity remains at 1,297 c.c., but the larger version went up to 1,598 c.c.

A GT derivative of the saloon, with more performance, has long been a standard catalogue model in the Cortina range. Also available, but only to special order, is a GT version of the estate car. This is not normally advertised and is not listed in the handbooks, spare parts lists, etc. Nevertheless, prospective purchasers should have no difficulty in obtaining one through a Ford dealer, although they are most unlikely to find one in stock.

Bodily, there has been no significant change to the Mark 2 Cortina estate car since its introduction in February 1967. The version tested last May was a 1500 Super, equipped with automatic transmission. Only the badges of this latest GT version distinguish it externally from the older car. Even the Super style stainless mouldings around the wheel arches are retained. Based on appearance, it could easily be dismissed as just another practical, but rather dull, work-horse. Nothing could be further from the truth—it is tremendous fun to drive, its performance and handling being out of all proportion to its basic utility character. Despite the fact that none of the practical advantages of an estate car have been sacrificed, only in terms of rear seat comfort does it fall short of the saloon.

In addition to the GT engine, it also has the associated close-ratio gearbox. Brakes and front suspension are to GT specification and radial-ply tyres on 4½J rims are standard, whereas they are an optional extra on non-GT models. In fact, apart from the estate car body and rear suspension, it is virtually identical to the GT saloon.

The most closely related Ford model we have tested is the 1600 E. The power trains are identical, and there is comparatively little difference in weight. The kerb weight of the estate car is 62lb greater, but this was reduced to 27lb in the "as tested" laden condition because of differences in the weights of the testers. This, not surprisingly, results in very similar performance figures. Acceleration times from 20 to 60 mph in top gear are identical at 21.9 sec and they are very nearly so from 40 to 80 mph (24.2 sec for the estate car, 24.1 sec for the 1600 E). Through the gears from rest, the times are also very similar. Up to 80 mph, those for the estate car are marginally slower, the difference in all probability being due to some lack of clutch "bite" experienced during the actual get-away. Above 80 mph, there is rather more discrepancy and the mean maximum is 2 mph down on that of the 1600E. This may well be due to normal variation between production engines. It could also be a function of the body shape, although generally we find estate cars faster than saloons. For all practical purposes, however, there is nothing to choose between the two.

Steady-speed fuel consumption figures are slightly better than those of the 1600E, but the overall, surprisingly, is worse (23.6 mpg, compared with 25.1 mpg). Although all our test staff usually participate in a full road test, it occasionally happens that a harder-than-average driver does a considerable proportion of the test mileage, with a consequent adverse effect on overall consumption. Ironically, the cars which are fun to drive suffer the most in this respect. Most owners would have no difficulty in equalling the 27 mpg typical figure we quoted for the 1600E.

None of the comparably priced estate cars we have tested can equal the performance of the Cortina GT. Based on the figures for the saloon counterpart, its nearest rival may well be the Victor 2000. Comparison with the now obsolete, 1498 c.c. engined Super, is superfluous, especially as the example tested by us was fitted with automatic transmission. It is interesting, however, to compare it with the two-litre V4-engined Corsair GT estate car which we tested in September 1966 (a conversion of the saloon by E. D. Abbot of Farnham). The Corsair's maximum speed was 5 mph lower

FORD CORTINA G.T. Estate Car (1,598 c.c.) Autocar road test Number 219

PERFORMANCE

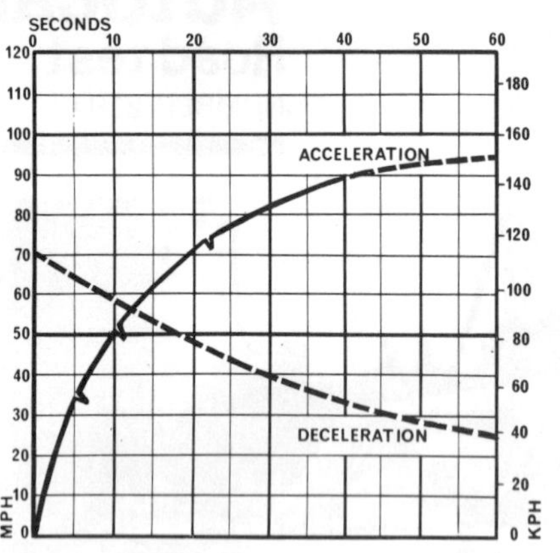

MAXIMUM SPEEDS

Gear	mph	kph	rpm
Top (mean)	96	155	5,620
(best)	97	156	5,670
3rd	74	119	6,000
2nd	51	83	6,000
1st	35	53	6,000

Standing ¼-mile 19.3 sec 70 mph
Standing kilometre 35.9 sec 87 mph

MOTORWAY CRUISING

Error (ind. speed at 70 mph)	72 mph
Engine (rpm at 70 mph)	4,090 rpm
(mean piston speed)	2,170 ft/min
Fuel (mpg at 70 mph)	29.2 mpg
Passing (50-70)	9.7 sec
Noise (per cent silent at 70 mph)	40 per cent

TIME IN SECONDS	0	4.5	6.8	9.6	14.2	19.3	27.1	41.5
TRUE SPEED MPH		30	40	50	60	70	80	90
INDICATED SPEED		32	42	52	62	72	82	90

Mileage recorder 3.0 per cent over-reading.

SPEED RANGE, GEAR RATIOS AND TIME IN SECONDS

mph	Top (3.90)	3rd (5.45)	2nd (7.84)	1st (11.59)
10-30	12.2	8.1	5.4	3.7
20-40	11.2	7.0	4.6	—
30-50	10.3	6.6	5.3	—
40-60	10.7	7.3	—	—
50-70	11.5	8.9	—	—
60-80	13.5	—	—	—
70-90	21.1	—	—	—

CONSUMPTION

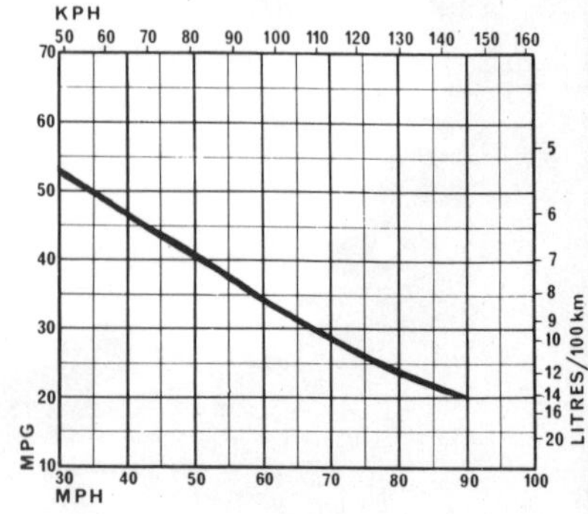

FUEL

(At constant speeds—mpg)

30 mph	52.7
40 mph	46.6
50 mph	40.4
60 mph	33.6
70 mph	29.2
80 mph	23.6
90 mph	19.5

Typical mpg . . 26 (10.9 litres/100km)
Calculated (DIN) mpg 26.6 (10.6 litres/100km)
Overall mpg . . 23.6 (12.0 litres/100km)
Grade of fuel: Premium, 4-star (min 97RM)

OIL

Miles per pint (SAE 20W/40) . . 700

HOW THE CAR COMPARES

Maximum Speed (mph)

70 80 90 100 110

Ford Cortina GT E.C.
Peugeot 404L Familiale E.C.
Triumph 2000 E.C.
Vauxhall Victor 2000 E.C.*
Volvo 145S E.C.

0-60 mph (sec)

30 20 10

Ford Cortina GT E.C.
Peugeot 404L Familiale E.C.
Triumph 2000 E.C.
Vauxhall Victor 2000 E.C.*
Volvo 145S E.C.

Standing Start ¼-mile (sec)

30 20 10

Ford Cortina GT E.C.
Peugeot 404L Familiale E.C.
Triumph 2000 E.C.
Vauxhall Victor 2000 E.C.*
Volvo 145S E.C.

MPG Overall

0 10 20

Ford Cortina GT E.C.
Peugeot 404L Familiale E.C.
Triumph 2000 E.C.
Vauxhall Victor 2000 E.C.*
Volvo 145S E.C.

*Saloon performance figures, estate car price

PRICES

Basic	**£842 0s 0d**
Purchase Tax	**£233 17s 9d**
Seat belts (pair)	**£8 6s 2d**
Total (in GB)	**£1,084 3s 11d**

TEST CONDITIONS Weather: Fine. Wind: 0-9 mph. Temperature: 24 deg C (76 deg F). Barometer: 29.6 in. Hg. Humidity: 46 per cent. Surfaces: Dry concrete and asphalt.

WEIGHT Kerb weight: 19cwt (2,126lb-965kg) (with oil, water and half-full fuel tank). Distribution, per cent: F, 51.3; R, 48.7. Laden as tested: 22.6cwt (2,534lb-1,150kg).

Test distance 1,510 miles.
Figures taken at 4,500 miles by our own staff at the Motor Industry Research Association proving ground at Nuneaton.

TURNING CIRCLES
Between kerbs: L, 30ft 2in.; R, 31ft 3in.
Between walls: L, 32ft 8in.; R, 33ft 9in.
Steering wheel turns, lock to lock: 4.7

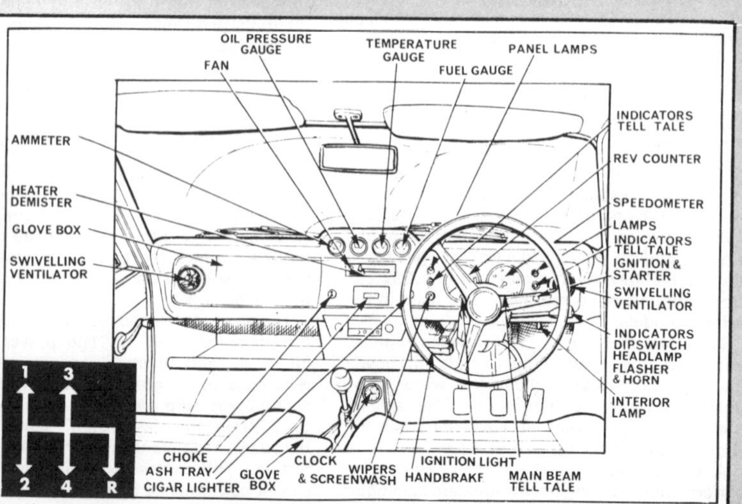

FORD CORTINA G.T. Estate Car (1,598 c.c.)

Autocar road test Number 2194

BRAKES

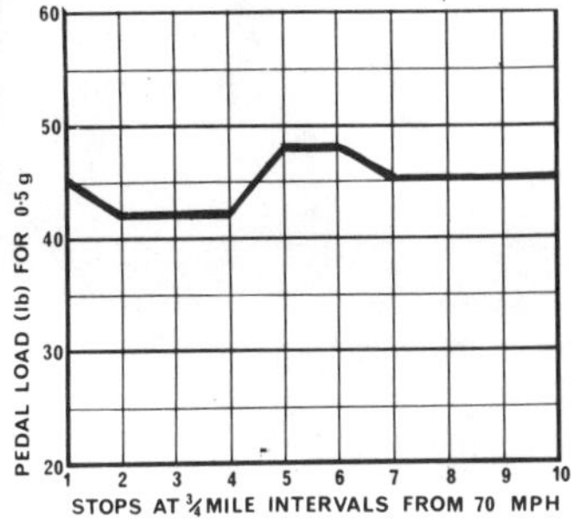

(from 30 mph in neutral)

Load	g	Distance
20lb	0.19	159ft
40lb	0.41	73ft
60lb	0.67	45ft
80lb	0.83	36ft
90lb	0.95	31.7ft
Handbrake	0.35	86ft

Max. Gradient 1 in 3
Clutch pedal: 32lb and 5.75in.

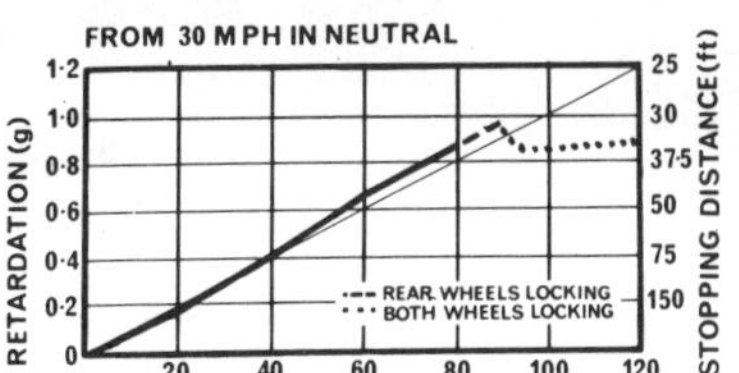

SPECIFICATION

FRONT ENGINE, REAR-WHEEL DRIVE

ENGINE

Cylinders	4, in line
Main bearings	5
Cooling system	Water; pump, fan and thermostat
Bore	80.98mm (3.19in.)
Stroke	77.62mm (3.06in.)
Displacement	1,598 c.c. (97.5 cu. in.)
Valve gear	Overhead, pushrods and rockers
Compression ratio	9.0-to-1; Min. octane rating: 97RM
Carburettor	One Weber 32 DFM compound twin choke
Fuel pump	AC mechanical
Oil filter	Fram or Tecalemit full flow, renewable element
Max. power	88 bhp (net) at 5,400 rpm
Max. torque	96 lb.ft. (net) at 3,600 rpm

TRANSMISSION

Clutch	Borg and Beck, diaphragm spring, 7.5 in. dia.
Gearbox	Four-speed, all-synchromesh
Gear ratios	Top 1 0
	Third 1.40
	Second 2.01
	First 2.98
	Reverse 3.32
Final drive	Hypoid bevel, 3.90-to-1

CHASSIS and BODY

Construction	Integral, with steel body

SUSPENSION

Front	Independent, MacPherson struts, coil springs, anti-roll bar, telescopic dampers
Rear	Live axle, half-elliptic leaf springs, lever arm dampers

STEERING

Steering	Burman recirculating ball
Wheel dia.	15.4in.

BRAKES

Make and type	Girling disc front, drum rear, no servo
Dimensions	F. 189.5in. dia.; R. 98.9in. dia.
Swept area	F. 189.5 sq.in.; R. 98.9 sq.in.; Total 288.4 sq.in. (257 sq.in./ton laden)

WHEELS

Type	Pressed steel disc, four-stud fixing, 4.5in. wide rim
Tyres—make	Radial-ply Dunlop, Firestone, Goodyear Michelin or Pirelli; Pirelli on test car
type	Cinturato radial-ply tubeless
size	165-13in.

EQUIPMENT

Battery	12 volt 38 Ah
Generator	Lucas C40L 22-amp d.c.
Headlamps	Lucas sealed filament, 90/120 watt (total)
Reversing lamp	Extra
Electric fuses	6
Screen wipers	Single speed, self-parking
Screen washer	Standard, manual plunger
Interior heater	Standard, air-blending type
Heated backlight	Available from Triplex
Safety belts	Extra, anchorages built in
Interior trim	Pvc seats, pvc headlining
Floor covering	Carpet
Starting handle	No provision
Jack	Screw pillar
Jacking points	Two each side, under sills
Windscreen	Zone toughened
Underbody protection	Phosphate treatment prior to painting

MAINTENANCE

Fuel tank	8 Imp. gallons (36.3 litres) (no reserve)
Cooling system	11.2 pints (including heater)
Engine sump	7.2 pints (4.1 litres) SAE 10W/50. Change oil every 6,000 miles. Change filter element every 6,000 miles
Gearbox	1.75 pints SAE 80EP. Change oil first 6,000 miles only
Final drive	2 pints SAE 90EP. Check level every 6,000 miles
Grease	No points
Tyre pressures	F. 24; R. 30 p.s.i. (all conditions)
Max. load	900lb (409kg)

PERFORMANCE DATA

Top gear mph per 1,000 rpm	17.1
Mean piston speed at max power	2,745 ft/min
Bhp per ton laden	77.7

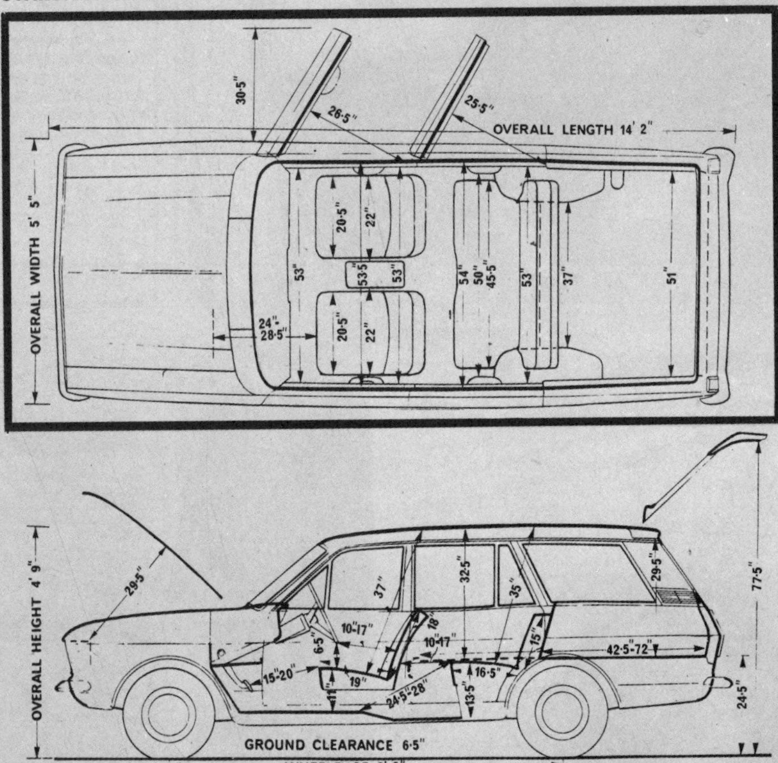

SCALE
0.3in. to 1ft
Cushions uncompressed

1

2

3

4

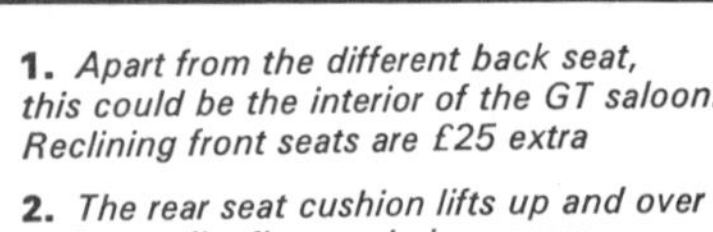

1. *Apart from the different back seat, this could be the interior of the GT saloon. Reclining front seats are £25 extra*

2. *The rear seat cushion lifts up and over to form a flat floor and give access to the jack*

3. *The GT facia incorporates rev counter and four supplementary dials sunk into the capping*

4. *Only the rear part of the load space is covered with rubber mats, but the floor is flat all the way to the front seats*

5. *GT engines have tubular branched manifolds and a twin-choke Weber carburettor*

6. *The one-piece tailgate lifts up and well out of the way for loading*

5

6

Only the GT badges on the rear quarters reveal that this is not an ordinary 1600 Super. Wide rimmed wheels and radial-ply tyres are standard

FORD CORTINA GT ESTATE . . .

and it took 0.4 sec longer to cover the standing ¼-mile. There is little to choose as far as fuel consumption is concerned. The load carrying capacity and general road behaviour of the Cortina are better, but its rear seat comfort is decidedly worse. The Corsair 2000 estate car, as it is now known, costs about £150 more than the Cortina GT estate car, but it now has the same engine specification as the 2000E, resulting in much improved performance.

During the warm weather with which we were blessed over the test period, starting was always immediate and drive-away flexibility very good. The oil pressure gauge takes 4-5sec to show a reading when the engine is first started after standing overnight, but this is not accompanied by any disturbing noises. Incidentally, no oil pressure warning light is fitted—such an item is a useful complement to the gauge, as it can attract the driver's attention more readily than a zero gauge reading.

Clutch and Brakes

Clutch pedal effort is low, but the clutch lacks "bite" during hurried starts, such as when taking performance figures. It also failed to cope with a re-start on the 1-in-3 test gradient but managed the 1-in-4. With the closer ratio GT gearbox, first gear is relatively high. Our earlier statement that none of the estate car qualities are sacrificed is not quite true when considering steep hill re-starting, but this is a penalty most GT owners will gladly suffer for the sake of the much better spaced ratios.

Brake pedal efforts are low enough without employing a servo, an 80lb load resulting in 0.83g retardation. The best mean figure recorded was only 0.95g, with an effort of 90lb. Behaviour during our fade test was even better than is usual for a GT Cortina, there being remarkably little variation in pedal effort and no misbehaviour of any kind. The excellence of the brakes when one is enjoying the car on twisty roads confirm these results. The handbrake produced a deceleration of 0.35g and could easily cope with the 1-in-3 test gradient.

For an estate car, particularly one with such impressive performance, the noise level is by no means excessive. The test car suffered from a bad "blow" from one of the silencer welded joints, but this was only evident from the outside, or when the car was idling in traffic with a window open. Road-generated noise is not loud and wind noise, although quite noticeable at the higher speeds, is not particularly distressing. Quite often, rear axle noise can be heard more in estate cars than in equivalent saloons but the Cortina is really very good. The body was surprisingly free from rattles, only the glove box lid offending in this respect. Above 80 mph or so, the general noise level rises fairly sharply. Even so, the car still has an unburstable feel about it, right up to the maximum of 96 mph.

The rev counter (part of the GT package) was over 8 per cent optimistic, so for our performance testing, we used an indicated 6,500 rpm for the gear change points, corresponding to a shade over a true 6,000 rpm. Many motorists tend to regard rev counters as more accurate than speedometers, but often this is not so, especially at high engine speeds. The speedometer on the test car, for instance, over-read consistently by only 2 mph up to and including 80 mph and was spot-on at 90 mph. The odometer over-read by 3 per cent.

All Cortina estate cars, including the GT, have the same rear suspension. This employs dual rate leaf springs, which enable heavy loads to be carried without the penalty of an uncomfortable ride when lightly laden. On the GT estate car, the front suspension is the same as that of the GT saloon. Whether by design or by accident, this results in a very nicely balanced vehicle. Previous Cortina estate cars we have tried, with their softer front suspension, lift their inside rear wheel rather easily when cornered exuberantly and oversteer markedly under these circumstances. The GT estate car is far better mannered. It understeers mildly up to quite high cornering forces, but when pushed really hard the rear wheels eventually slide out in a progressive and controllable fashion. Stability is not adversely affected by load, although we never carried anything really heavy.

The steering is light and has plenty of feel. Response is sufficiently rapid, the ratio feeling considerably higher than the figure of 4.7 turns from lock to lock suggests, because the turning circle is remarkably compact at a maximum of 31ft 3in. between kerbs. Straight line stability compares favourably with other Cortina derivatives, but we didn't experience any severe cross-wind conditions during our test.

The ride, although not outstanding, seems better than on the lower performance version. Front seat occupants have the advantage of well damped seats. Those on the test car had reclining backrests, available at extra cost. They feel a trifle hard, but provide reasonable location during hard cornering and have a good range of adjustment. The rear seat, however, leaves a great deal to be desired. It has insufficient damping to cope with the rather lively ride, a situation magnified by the fact that the backrest is so shallow that its top edge is actually below the average adult's shoulder blades. This low backrest is one of the penalties resulting from a seat folding arrangement designed to provide a completely uninterrupted floor length, right up to the front seats. Another is the absence of a barrier to restrain the load in the event of a sudden stop. Many will consider these disadvantages outweighed by the fact that the Cortina is a particularly spacious load carrier in relation to its external size, and this flat floor is long enough to make a full size bed for camping.

Interior Trim

Apart from the individual front seats, the additional instruments and the manual transmission, the interior is the same as that of the Super tested last year. None of the features criticized in that report have been changed. It seems a pity that the enamelled surfaces of the additional floor space formed by the folded seat are not provided with similar protective mats to the excellent moulded rubber one which covers the rear.

Since an estate car lacks a saloon's "private" boot, the additional concealed stowage space which the GT's centre console provides for personal items is useful. Neither this nor the glove box can be locked, however.

One detail which has been changed during the past year is the method of controlling the cold air vents, each "eyeball" now incorporating a knob for opening and closing a butterfly valve. Neither was capable of completely shutting off the supply of cold air. Furthermore, the left hand one persistently refused to remain open at speeds over 70 mph.

The Cortina GT estate car is by no means cheap. In the form tested, it retails for a total of nearly £1,137. Extras fitted included the reclining seats, a metallic paint finish, a non-push-button radio and the mandatory seat belts, in this instance of the excellent inertia reel type. It is generally very well constructed and finished, only a few details deserving criticism. The most obvious of these are the shoddy coat hooks, retained by pop-rivets, both of which were loose. Bad fitting made the left hand rear door difficult to open and we think that a car costing this much deserves at least a self-supporting bonnet prop and an internal release. These are trifling criticisms, being insignificant in comparison with the shortcomings of the rear seat. The latter feature is most unfortunate since this is very much a family car and those in the back are not going to be happy. There is little doubt, however, that those who can accept this limitation will be delighted with this dynamic and versatile vehicle.

62

CAR LIFE ROAD TEST

the ageless CORTINA

We expected thrills with Ford's Model "C" in GT getup—instead fun turned to respect when the car was under stress

IT'S A RESPONSIBLE car, the Cortina GT, ruggedly built with some sound engineering behind it. It has adequate power, and very sensible handling. In fact, that should capsule it quite well: sensible and adequate. But if Ford of England meant "GT" when it named this Cortina, then it doesn't pass. Probably we were expecting a mini-Ponycar, when in fact we got a well arranged version of an economy car. CAR LIFE wasn't disappointed, it was just that we embarked on the test expecting to have fun and came back serious and respectful. Beef steak instead of cake and ice cream.

The GT package consists of an interior/exterior trim group (tach, instruments, console), a handling group (stiffer springs, shocks and radial tires), and the cam/carb/compression treatment along with a close-ratio four-speed gearbox. All this is sprinkled liberally over the "Model C" supposedly to turn it into a family sports sedan. The result is likeable, but a sports car it isn't. Do not expect to blow off a Camaro at the stoplight GP, or dust a Corvair on the canyon road. You can expect to eat up any VW, Datsun or Toyota sedan in town or on the road of his choosing.

The Cortina GT's strong point is its handling. There's nothing really novel in the suspension design—the engineers have just tied everything down well with medium-stiff springs, decent shocks and logical locating members. (The rear axle even has radius rods.) First impression of the handling is a bit disconcerting. The high center of gravity makes the body lean over quite markedly when first entering a corner, giving the driver that oh-my-gosh-I'm-gonna-tip-over feeling. Careening around the handling course with the photographer (a fearless sky diver and stunt pilot) as passenger con-

NEUTRAL STEERING and lifting of inside rear wheel could be induced by very spirited cornering maneuvers.

firmed this when he kept wanting to bail out. However, once it assumes this keel-over attitude, it hunkers down and becomes a very predictable machine, negotiating the turns with an understeering to neutral steering characteristic. Oversteer was impossible to obtain. The inside rear wheel would unload and spin when throttle was applied making it impossible to push the tail out. Just as well, too, as the average motorist can rarely cope with oversteer anyway.

As the driver's confidence builds up, it begins to be fun. The tendency is to charge into a corner, assuming nice neutral drift early and continuing it with the throttle while backing off the steering wheel exiting the turn, coming onto the straight with full throttle already on. It feels like the old Trans-Am Lotus-Cortinas used to look: The car flopping over to seemingly full spring travel at the corner entrance before driving under someone on the inside at the corner exit. Oversteer was induced only once during the handling tests, by going into a corner at 75 mph (but the tester swore never to repeat it again, even for money).

Increased spring and shock rates along with the sturdy chassis construction made the GT behave exceedingly well on rough surfaces, never pitching or bouncing regardless of bump amplitude or frequency. We got the impression it would be a good back-road car.

Braking wasn't bad, but should have been better. The maximum deceleration of 25 ft./sec./sec. was obtained only after much practice of pedal modulation trying to keep from locking up the rear wheels, something not available to a driver in a normal panic situation. The classic panic stab of the brake pedal inevitably brought on rear-wheel lock up and loss of direction stability.

The overall construction was impressive. Nothing rattled, everything looked and felt sturdy without being overly heavy or stiff. The only things that gave the illusion of looseness was the initial cornering lean and a slightly whispy feeling gear lever. The gating was clean and positive, but the longer-than-usual throw and willowy lever seemed out of character with the rest of the car. Also the gate was the same as for the right-hand drive model, which meant it was tilted slightly to the right and the second to third shift required concentration to keep from snagging first. Fortunately, the testers managed to get the knack of this before getting to the dragstrip and the 6000-rpm snap shifts.

The interior was above average for an economy import, but then that is part of the bargain. Remember, it is a GT. Instrument panel was very good, having a big speedometer and tachometer side by side in front of the driver and the engine instruments in a neat panel directly below the rearview mirror. It probably was not as neat as having them clustered somewhere nearer the driver.

Worst part was the lock mechanism on the seats. Not the knob itself—*that* works. But the seat is one piece. The back doesn't tip forward, the whole seat comes up and forward from hinges on the front of the seat frame. Clearing a path for entry to the rear seat requires that the front seat be emptied before a passenger can enter the back. Then when the seat is back in place, the belt is stuck underneath. Sometimes it's the floor strap, sometimes the roof, sometimes both.

Always interesting in any Ford of England is that swinging engine. The Cortina engine is to England what

PHOTOS BY PAUL E. HANSEN

DECELEROMETER on windshield recorded minus 25 ft./sec./sec. before rear wheels locked.

CORTINA GT

continued

Chevy's small V-8 is to the United States. It has been the basis for uncountable racing engines, many of which have washed ashore in the States, where savvy racers are quick to notice its potential also. It is hard

1968 CORTINA

FORD GT

DIMENSIONS

Wheelbase, in.	98.0
Track, f/r, in.	52.5/51.0
Overall length, in.	168.0
width	64.9
height	55.0
Front seat hip room, in.	2 x 22
shoulder room	53.0
head room	37.1
pedal-seatback, max.	40.0
Rear seat hip room, in.	52.0
shoulder room	53.0
leg room	35.3
head room	36.8
Door opening width, in.	40.0
Trunk liftover height, in.	31.0

PRICES

List, FOB factory	$2384
Equipped as tested	$2384
Options included: None.	

CAPACITIES

No. of passengers	4
Luggage space, cu. ft.	21.0
Fuel tank, gal.	12.0
Crankcase, qt.	3.12
Transmission/dif., pt.	1.75
Radiator coolant, qt.	6.6

CHASSIS/SUSPENSION

Frame type: Unitized.	
Front suspension type: Independent by MacPherson strut, combined lower control arm/anti-roll bar, coil springs & telescopic shock absorbers.	
ride rate at wheel, lb./in.	139
antiroll bar dia., in.	0.810
Rear suspension type: Hotchkiss drive, solid axle, semi-elliptic leaf springs, telescopic shock absorbers, torque control arms.	
ride rate at wheel, lb./in.	85
Steering system: Recirculating ball.	
overall ratio	16.4:1
turns, lock to lock	4.5
turning circle, ft. curb-curb	30.0
Curb weight, lb.	2030
Test weight	2320
Distribution (driver), % f/r	55/45

BRAKES

Type: Disc front, drum rear self adjusting.	
Front rotor, dia. in.	9.625
Rear drum, dia.	9.00
total swept area, sq. in.	285.6
Power assist: None.	

WHEELS/TIRES

Wheel rim size	13 x 4.5J
optional size	none
bolt no./circle dia. in.	4/4.5
Tires: Goodyear radials.	
size	165-13
normal inflation, psi f/r	24/28
Capacity @ psi	3700

ENGINE

Type, no. of cyl.	ohv IL4
Bore x stroke, in.	3.189 x 3.056
Displacement, cu. in.	97.6
Compression ratio	9.6:1
Fuel required	premium
Rated bhp @ rpm	89 @ 5500
equivalent mph	92
Rated torque @ rpm	102.5 @ 4000
equivalent mph	67
Carburetion: 1x2 Weber.	
throttle dia., pri/sec.	1.1/1.42
Valve train: Mechanical lifters, push rods and overhead rocker arms.	
cam timing deg., int./exh.	27-65/65-27
duration, int./exh.	272/272
Exhaust system: Four-branch tubular headers, reverse-flow muffler.	
pipe dia., exh./tail	1.625/1.625
Normal oil press. @ rpm	40 @ 3000
Electrical supply, V./amp	12/30
Battery, plates/amp. hr.	13/38

DRIVE TRAIN

Clutch type: Single plate, diaphragm spring.	
dia., in.	7.54
Transmission type: All synchro four-speed.	
Gear ratio 4th (1.00:1) overall	3.90:1
3rd (1.397:1)	5.45:1
2nd (2.01:1)	7.84:1
1st (2.972:1)	11.60:1
Shift lever location: Console.	
Differential type: Semi-floating hypoid.	
axle ratio	3.90:1

SUPPLE SUSPENSION, good tire adhesion allowed "scary" body roll, but predictability and cornering power were above average.

to put a finger on just what one thing makes it so good. The sturdy four-cylinder, five-main-bearing block is a virtual pillar of strength. Short stroke and very light valve gear combine to make it one of the revviest engines on the market. In 1967, Ford of England introduced the crossflow cylinder head with the combustion chamber in the piston to improve already good breathing and combustion efficiency. Yet this seemingly elaborate engine is actually simpler and has more interchangeability than any "modern" engine on the market, except possibly the Chevy.

Also in '67, 0.18-in. stroke was added to the crank bringing the engine

CAR LIFE ROAD TEST

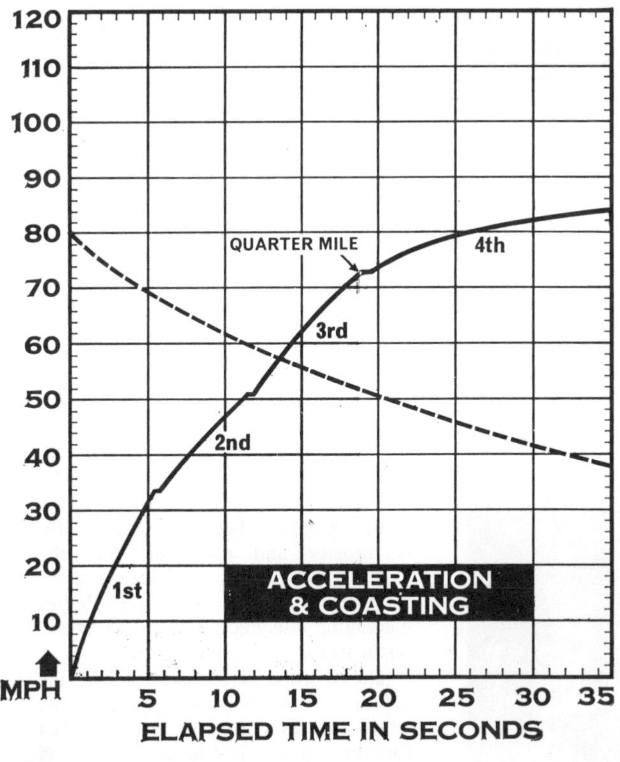

CALCULATED DATA

Lb./bhp (test weight)..........26.1
Cu. ft./ton mile................86.2
Mph/1000 rpm (high gear)......16.8
Engine revs/mile (60 mph).....3570
Piston travel, ft./mile..........1817
CAR LIFE wear index..........64.9
Frontal area, sq. ft.............19.6

SPEEDOMETER ERROR

30 mph, actual.................31.2
40 mph.......................40.8
50 mph.......................50.2
60 mph.......................59.6
70 mph.......................69.2
80 mph.......................79.9
90 mph.......................88.1

MAINTENANCE

Engine oil, miles/days.........6000
oil filter, miles/days.........6000
Chassis lubrication, miles.......6000
Antismog servicing, type/miles replace/12,000
Air cleaner, miles.............6000
Spark plugs: Autolite AG-22A.
gap, (in.)....................0.023
Basic timing, deg./rpm..10BTDC/600
max. cent. adv., deg./rpm...........11-13/3000
max. vac. adv., deg./in. Hg.......... 5.5-8.5/10
Ignition point gap, in...........0.025
cam dwell angle, deg...........39
arm tension, oz..............7-21
Tappet clearance, int./exh..............0.010/0.017
Fuel pressure at idle, psi.....3.5-5.0
Radiator cap relief press., psi.....13

PERFORMANCE

Top speed (6100), mph..........102
Test shift points (rpm) @ mph
3rd to 4th (6000)..............73
2nd to 3rd (6000)..............51
1st to 2nd (6000)..............34

ACCELERATION

0-30 mph, sec...................4.9
0-40 mph.......................7.8
0-50 mph......................11.3
0-60 mph......................14.9
0-70 mph......................17.6
0-80 mph......................26.5
Standing ¼-mile, sec...........18.8
speed at end, mph.......72.5
Passing, 30-70 mph, sec.........13.1

BRAKING

Max. deceleration rate from 80 mph ft./sec./sec...................25
No. of stops from 80 mph (60-sec. intervals) before 20% loss in deceleration rate...........8-no loss
Control loss? Yes.
Overall brake performance.......fair

FUEL CONSUMPTION

Test conditions, mpg.............22
Normal cond., mpg...........26-28
Cruising range, miles.......310-340

CORTINA GT

continued

LOW TORQUE and high adhesion required liberal use of throttle for fast starts.

NEW crossflow head with large valves and combustion chamber in piston crown achieve high gas flow, good power and emission control potential.

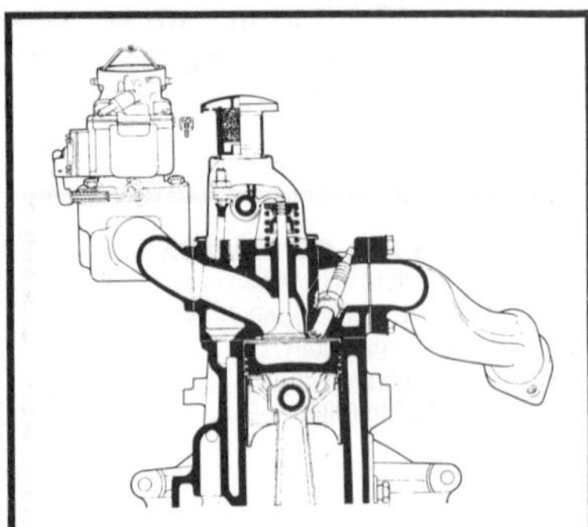

SUPERB engine design and quality workmanship were Cortina virtues. Exhaust headers are standard on GT, smog pump is required for U.S.

HANDSOME INTERIOR is also functional and comfortable. Instruments include oil pressure, water temperature, ammeter, large tach. Pedals and gearshift are well placed.

up to 98 cu. in., and nearly squaring the engine (3.19 bore by 3.06 stroke). A slightly hotter cam, a progressive two-barrel Weber carburetor, 9.6:1 compression ratio, and a set of fabricated headers bring the engine up to 89 bhp at 5500 rpm. Though this isn't exactly blood and thunder when propelling an even ton of automobile, it is entirely adequate. Like everything else about the car, the power is commensurate with the braking and handling. Like we said, the car is responsible.

It is surprising to find a smog pump on this ultra-modern engine. Most U.S. manufacturers have found ways around them via better mixture control and optimum spark advance. This is especially so when considering the low area-to-volume ratio of Ford's combustion chamber design (good for smog control). We suspect that it was an economy move. Ford markets this car, and the smaller Escort, all over Europe where smog control is not yet required. It is far simpler to make jillions of the same set-ups (which are rather performance oriented—slightly richer mixtures, more emissions) and then add the cheap air pump for the American market.

Gearbox ratios seemed about right, the close ratios handy for maneuvering in traffic and playing games on mountain roads. Low gear was rather high for the camming of the engine and it took more than normal revs to make clean get-aways. It wasn't nerve wracking, but it did take some getting used to. It felt like it was willing to cruise all day at any speed one felt like pushing it, all the way up to the 6000-rpm redline, hauling along freeways with the best of them, with not so much as a single vibration—unique for a four-cylinder. We got the impression that the optional 4.13:1 rear axle ratio would be better, helping out the not-too-impressive acceleration and lopping off some of that unneeded top end. This would hurt the fuel economy, which is nothing to write home about anyway, but it would be worth the sacrifice.

Styling reflects the overall personality of the GT—sensible, adequate, and responsible. The interior was probably better than the outside. One staffer, staring out the window (as usual) made a startling discovery: The Cortina is nearly the same size as the Mustang parked beside it. Only slightly shorter and narrower. Yet the Cortina has a roomier and more comfortable interior. Simply stated: better packaging efficiency at the cost of exterior styling.

The Cortina is an ageless car, the old ones never look old, the new one never startles you. They seem to last forever—they're just around. ■

GROUP TEST

'Motor's' test team go driving in convoy to try competitive cars under identical conditions

No. 11

Fast medium saloons

- Ford Capri 1600 GT
- Fiat 125
- Ford Cortina 1600E
- Vauxhall Viva GT
- Triumph Vitesse 2-litre
- Ford Escort TC

Bonnets open different ways while Motor's convoy stops for vital fluids. From bottom to top, Vauxhall Viva GT, Ford Capri GT, Fiat 125, Ford 1600E, Triumph Vitesse and Ford Escort Twin Cam.

THIS GROUP is really a collection of wolves in sheep's clothing—cars that are faster and more entertaining than their similar looking pottering counterparts; if you accept that the Vitesse is virtually a very hot Herald, then the Viva GT, the Capri 1600 GT, the 1600E, and the Escort TC also share bodies with unpretentious family saloons. Even the Fiat 125 owes much of its design to the 124 but, in any case, its general road behaviour and performance certainly justify its inclusion in a group which might loosely be summed up as high performance cars for those who can't afford to sacrifice space and probably don't want to afford the insurance companies' frowns of disapproval—sports cars without tears, in short.

Accepting that such owners would like their transport to be capable of around 100 m.p.h. they will have to pay between £1,000 and £1,200 for the accompanying standards of refinement. For almost the same performance you can keep the Vitesse under £1,000 by having it without an overdrive, but for motorway work even at 70 m.p.h. this would be a bad idea; 1600E performance can, of course, be retained with a straight Cortina GT for under £1,000, but then you don't get the wider wheels nor the Cortina Lotus suspension, both worth having, let alone the much more comfortable and plusher interior. Another obvious candidate we haven't included is the Cortina Lotus; we preferred to split its engine and body into the Escort TC and the 1600E, so if you want to know what we think of the Cortina Lotus you can either wait for a later group or, with qualifications, take the relevant comments from its two cousins.

If our selection looks like a Ford benefit, this is because Ford have a blanket coverage of this market with cars designed specifically for it, and promoted accordingly. Their three contributions have widely differing characters but they all belong here.

Continued on the next page

The cars

Ford Capri 1600GT—XLR £1,121
Integral steel body; independent Macpherson strut front suspension; live rear axle with leaf springs and radius arms; rack-and-pinion steering; disc/drum brakes; 1,599 c.c. pushrod o.h.v. four with Heron head engine driving back wheels through four-speed box; Dunlop SP68 tyres.

Fiat 125 £1,125
Integral steel body; independent wishbone front suspension; live rear axle with leaf springs and radius arms; worm-and-roller steering; disc brakes all round; 1,608 c.c. twin overhead camshaft four mounted at front driving rear wheels through four-speed gearbox; Pirelli Cinturato tyres.

Ford Cortina 1600E £1,097
Integral steel body; independent Macpherson strut front suspension; live rear axle with leaf springs and radius arms; recirculating ball steering; disc/drum brakes; 1,599 c.c. pushrod o.h.v. four with Heron head engine driving rear wheels through four-speed gearbox; Goodyear G800 tyres.

Vauxhall Viva GT £1,088
Integral steel body; independent wishbone front suspension; live rear axle with four link location and coil springs; rack-and-pinion steering; disc/drum brakes; 1,975 c.c. single overhead camshaft four driving rear wheels through four-speed gearbox; Goodyear Grand Prix G800 tyres.

Triumph Vitesse 2-litre £1,034 (inc. overdrive)
Separate steel chassis with steel body; independent wishbone front suspension; independent rear suspension with transverse upper leaf spring and lower wishbone; rack-and-pinion steering; disc/drum brakes; 1,998 c.c. pushrod o.h.v. six driving rear wheels through four-speed gearbox and optional overdrive; Goodyear G800 tyres.

Ford Escort TC £1,195
Integral steel body; independent Macpherson strut front suspension; live rear axle with leaf springs and radius arms; rack-and-pinion steering; disc/drum brakes; 1,558 c.c. twin overhead camshaft four driving rear wheels through four-speed gearbox; India Autoband tyres.

Might almost be Makinen on the RAC Rally! The Escort displays astonishing controllability while the Vitesse frowns disapprovingly, although it can quite easily be made to do the same on such a surface.

Group test No. 11

continued

Full road tests have been published on all these cars, and those on the Capri GT, Fiat 125 and Triumph Vitesse are still valid; the other three have acquired minor improvements since our original full tests. On the 1600E the facia has been restyled with all the smaller instruments sitting in the wood rather than in a little bulge on top of it; this pushed the heater controls down to where the ashtray used to be. All the knobs conform to American standards. A more complete console still houses the clock but is restyled to allow a pull-up handbrake rather than the unpleasant umbrella type. Outside, FORD is written boldly on the bonnet top, and the rear panel is matt black with chrome strips above and below. In all, some worthwhile changes.

The Escort TC, too, has benefited from internal safety changes but is otherwise unchanged although this car, a genuine production one, was quieter and more stable at speed than the original test car, a handbuilt prototype which probably had too much negative camber on the front wheels.

Improvements to the Viva GT have changed the driveability quite appreciably; a lower steering column makes the driving position much better and some of the switches have been placed on the console *a la* Ventora where they can more easily be reached by a belted driver.

Performance

The engines in this group present quite an interesting array. The most developed unit in terms of power/litre is the Escort TC with twin overhead camshafts; the Fiat 125 also has twin overhead cams, belt driven; belt drive is also used for the single overhead camshaft Viva engine. Pushrods operate the Capri and 1600E BIP engines (which are identical in specification) and the Vitesse unit, the only six-cylinder in the group.

In sheer performance the Escort TC leads the field comfortably; it has the smallest engine but the highest claimed power and the lightest overall weight by some 2 cwt.; this partly explains why it is 2 seconds quicker than the next fastest, the Viva GT, to 60 m.p.h. (1½ to 50). But then 0-60 m.p.h. in 8.7 s. is *very* fast; the rest take over 10 s., with the Viva, Vitesse, 1600E and Fiat 125 covered between 10.7 and 11.9 s. and the Capri trailing at 12.7 s., losing out slightly to the similarly engined 1600E (surprisingly, of identical weight) by virtue of higher gearing and probably better resistance to wheelspin during standing starts. A study of flexibility begins to make the story a little clearer: although the Escort is very much a top end performer, it is still faster than the other two Fords from 30-50 m.p.h. but because the whole thing is so taut and harsh below 30 in top it came pretty low in flexibility marks. The Viva GT was rather similar; it would pull strongly from 20 in top but not smoothly until about 26 m.p.h. The Fiat and the two other Fords pulled gently but

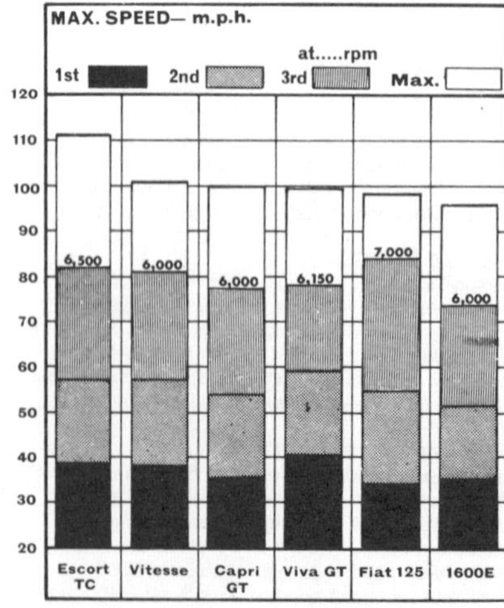

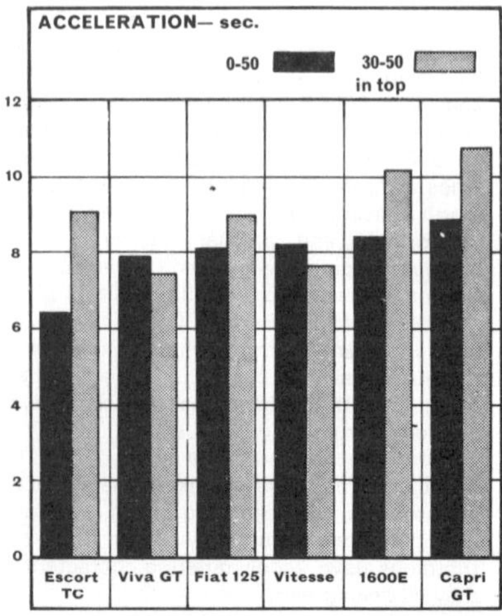

Ford's middleweight Executive shows less of the hip kink than Vauxhall's boy racer, which is faster but less refined.

quietly from about 25 with the Capri feeling that much happier than the others. The Vitesse was outstanding for flexibility—it pulls with all the smoothness that one expects from a six, and quickly too, although the Viva matches it against the stopwatch. It is still smooth and quiet at the top end of its rev scale, but matched in both by the wonderfully refined engine of the Fiat 125 which sounds and feels unburstable.

The engines of the Capri, Viva and Escort were all fairly smooth but didn't feel as unstrained as those of the Fiat and Vitesse; the Viva and Escort were noisy, too, under hard acceleration, the Escort particularly so in a rather unpleasant harsh fashion, quite unlike the sewing machine quality of the same engine in an Elan, for instance. The 1600E engine sounds rather the same over 4,500 r.p.m. but is fairly quiet up to this; the same engine in the Capri seems much better insulated throughout the range.

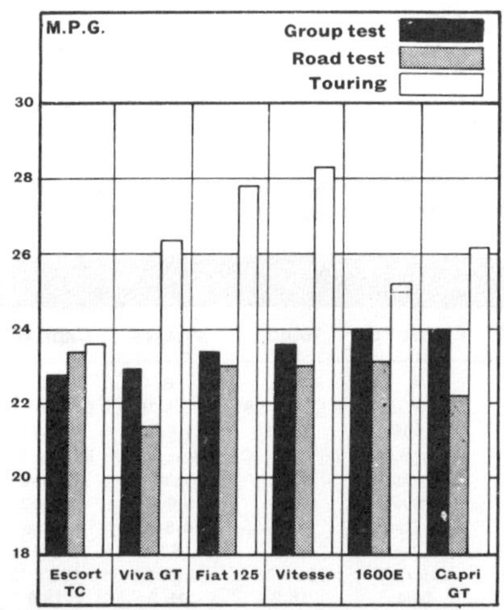

When sailing along at 70 m.p.h. all these 100 m.p.h. cars are well within themselves. If wind disturbance was as peaceful as the engine, the Vitesse would be a very quiet car. As it is the Capri was reckoned to be the most effortless, with the lowest wind noise of the group, very little road noise and a fairly quiet engine. Despite having the lowest overall gearing, the Fiat 125 was also very good; this was also the heaviest car in the group so perhaps its weight has been used to good sound-deadening effect. The other cars were noticeably noisier, with the Escort losing in engine roar what it gains in low wind noise; none of them are objectionable at this speed although another 15 m.p.h. for Continental cruising would probably bring some to a tiring level.

Running costs

Such a remarkable similarity in fuel consumptions on our 600-mile test run is most unusual; they were all within 0.7 m.p.g. of the mean 23.4 m.p.g. The Escort was marginally heaviest on fuel at 22.7 m.p.g. despite the fact that its extra potential allowed it to be driven slightly more gently than the heavier, less powerful cars in the same convoy—there is virtually no difference between its Road Test Overall and Touring consumptions. The overdrive on the Vitesse, which can usually save around 10% on fuel, failed to give the car an advantage over the others, probably because it was rarely used in the Welsh mountains. Both the Capri and 1600E returned better figures than for the Road Test, at 24.0 m.p.g.

These figures seem fairly heavy. None of the touring consumptions top 30 m.p.g., the Vitesse probably having the best chance of being significantly more economical in other hands. All of them run on 4-star fuel and the annual petrol cost worked out over 10,000 miles at 6s. 4d. is as follows:

	galls	£
Ford Capri 1600GT	417	132
Ford Cortina 1600E	417	132
Triumph Vitesse 2-litre	424	134
Fiat 125	428	135½
Vauxhall Viva GT	437	138
Ford Escort Twin-cam	441	139½

This represents a very small variation; compared with the previous sporting group where the most economical, the Fiat 850S Coupé, cost £92 and the thirstiest, the Escort 1300GT, £115, so you can save quite a lot in a year by sacrificing a little performance and probably not lose much in average speeds.

Based on the group rating system, insurance premiums seem to conform roughly to accelerative ability; the 1600E and Capri GT in group 4 will be about £25 a year cheaper to insure than the group 6 Viva and Escort (assuming no bonuses in a home counties area) and the Vitesse and Fiat 125 half way between in group 5.

All the cars can be serviced at well established networks including the one foreigner, the Fiat. Five of them need servicing every 6,000 miles, and the sixth, the Escort Twin-Cam, needs it far more frequently at 2,500-mile intervals; this is not just because it has a slightly more highly tuned engine, but the whole level of tune—engine, suspension, drive line—is far tauter than that of the average production car and is possibly there-

Continued on the next page

Group test No 11

continued

fore more prone to minor loosening of its joints. Totted up over a year, this would mean quite a heavy addition to the running costs.

Transmission

Ford gearboxes have become renowned for their excellence throughout the Escort, Cortina and Corsair ranges. Both Capri and 1600E use the same box, with its super-light and "well oiled", precise lever movement; they have the same clutches, too, but the better insulated drive line on the Capri makes this take-up less sharp than on the 1600E; both these and the Escort have the same well chosen ratios, but the Escort has a Corsair gearbox with a different reverse gear position and a three-rail selector shift instead of the other's single rail. The result of this is somehow to narrow the movement across the gate which is excellent once you are used to it although it is easy to select third instead of first when stationary. The Escort, too, has a much more rigid drive line and the clutch take-up is unpleasantly sharp.

We thought the Ford gearchange was the best in production, but a solid vote puts the Fiat just ahead, largely because it was matched by a more progressive and softer clutch action. Its intermediate ratios are lower, but the engine's free revability does a lot to close the 3-4 gap. Ratios on both the Vitesse and Viva are excellent and would suit far less docile pieces of machinery than these, both being capable of well over 50 m.p.h. in second. Unfortunately, the changes on both these cars are a bit sticky and notchy, and the gears whine quite loudly in the indirects which, for the Vitesse, means overdrive as well. The Fiat 125 had a steady-speed back axle whine but the Fords were almost silent.

Handling

This group provided a contrast between the current vogue for safe but uninspiring understeer, represented by the Viva and to a lesser extent by the Capri and the Fiat, versus controllable oversteer for the Escort and 1600E, with the Vitesse coming in between. Since they all have rear drive, the understeerers can be provoked to oversteer in the wet but they need a lot of surplus power to do so, particularly the Capri. A requirement for understeering cars is steering through which you can feel front end breakaway, while oversteerers can get away with steering which is just quick.

It needs a really slippery surface to get the Capri's tail this far out of line—this is the GT with XLR options. The Fiat 125 looks boxy by contrast but the shape is more practical.

The Capri steering suffers from lack of feel; it seems to remain constant in weight regardless of cornering forces but in other respects it was well praised for accuracy and freedom from kickback. The roadholding was good on all surfaces, and we really felt it quite impossible to lose it at sensible speeds. The tyres fitted were Dunlop SP68. With such a high limit of adhesion it is always safe and virtually fool proof.

Although it is easier to provoke the Escort's limit, we still rated its roadholding as first class. This one, on India Autobands, was very much better on slippery surfaces than our earlier Road Test car which suffered from too much negative camber; the steering is very quick and provides good feel. Everything about the car is so taut and alert that the Escort excels in responsiveness by comparison with the rest.

Very much in the same roadholding bracket as the Escort—but much less responsive—is the Viva GT on its low profile radials (Goodyear Grand Prix G800); the tail can be provoked on Hereford's muddy roads but it is extremely adhesive nonetheless, and the normally predictable understeer doesn't become excesive, making the car run wide, until you are going very quickly indeed. Some people thought the steering was overdamped and masked useful feel.

The roadholding of these three cars is very much on a par with and slightly better than the rest; the Vitesse came next by a short head. Its roadholding is much improved by the new independent rear suspension; live axles used to be prone to hopping on bumpy corners, but now they all seem to be so well developed that an independent rear has less advantage—except perhaps in snow—other than for ride improvement. Both grip and ride on the Vitesse rate in the middle of the group. Its handling is just about neutral but it can either under or oversteer according to throttle opening and cornering speed, and it will tuck in gently if you lift off in a corner. The steering is informative but not free from kickback, which drops its rating slightly.

Perhaps it reflects the style of our driving that the two cars whose handling we all liked most were the two with the lowest ratings for roadholding, the Fiat on Pirelli Cinturatos and the 1600E on Goodyear G800s. Both cars have quite reasonable limits of adhesion but they slide so politely, and can be caught so smoothly that you can indulge in high-slip cornering, if you know how, without danger of coming to grief. Reassuring. Pushed hard the Cortina will oversteer whereas the Fiat under or oversteers depending on several things; with either, you can chuck them in and go through the corner on a predictable and smooth line on the verge of sliding—a sort of driver-induced drift.

In all this it is worth remembering that cornering force goes up as the square of the speed, so that a 10% increase in forces on the tyres is produced from an increase in speed of only 5%, so the lesser cars never drop noticeably behind on give-and-take roads—it would be more noticeable on an Alpine pass, perhaps.

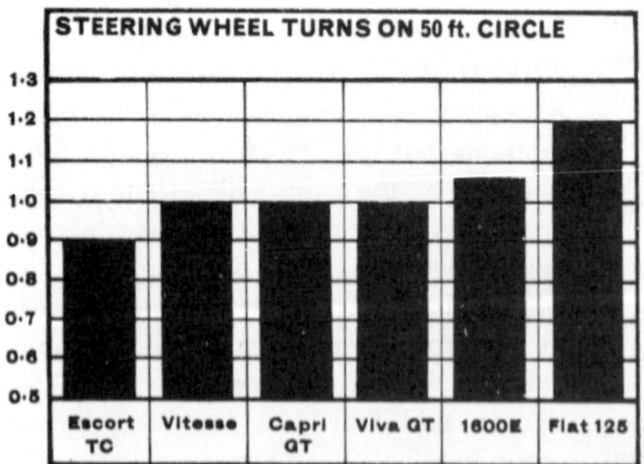

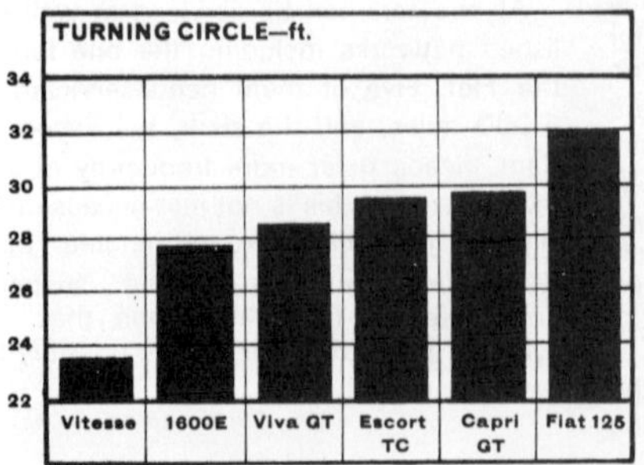

Specification

	Viva GT	Fiat 125	Escort TC	1600E	Vitesse	Capri GT
Cylinders	4	4	4	4	6	4
Bore and stroke (mm.)	95.2 x 69.2	80 x 80	82.5 x 72.7	81.0 x 77.6	74.7 x 76.0	81.0 x 77.6
Capacity (c.c.)	1,975	1,608	1,558	1,599	1,998	1,599
Net b.h.p. at r.p.m.	104 at 5,600	90 at 5,600	109 at 6,000	88 at 5,400	104 at 5,300	88 at 5,400
Brakes	Disc/drum	Disc/disc	Disc/drum	Disc/drum	Disc/drum	Disc/drum
Service internal (miles)	6,000	6,000	2,500	6,000	6,000	6,000
Fuel grade	4-Star	4-Star	4-Star	4-Star	4-Star	4-Star
Insurance rating (AOA/Lloyds)	6/6	4/5	6/O.A.	4/4	5/5	4/4
Tyre size	155 x 13	175 x 13	165 x 13	165 x 13	155 x 13	165 x 13
Weight (cwt.)	18.2	19.4	16.4	18.2	18.9	18.2
M.p.h./1,000 r.p.m. in top	17.0	16.7	17.6	17.2	17.0/21.2	17.9

Brakes

They all worked well. We didn't really have any strong feelings against any of them; perhaps the Vitesse brakes were a little spongy, and the Escort's overservoed, which sometimes made heel-and-toe changes a bit jerky, but these were minority opinions. With the repositioning of the 1969 1600E handbrake, all cars have them in the convenient centre console position. Worthy of note in this section is the astonishing retardation of the Viva on greasy roads, again thanks to its tyres.

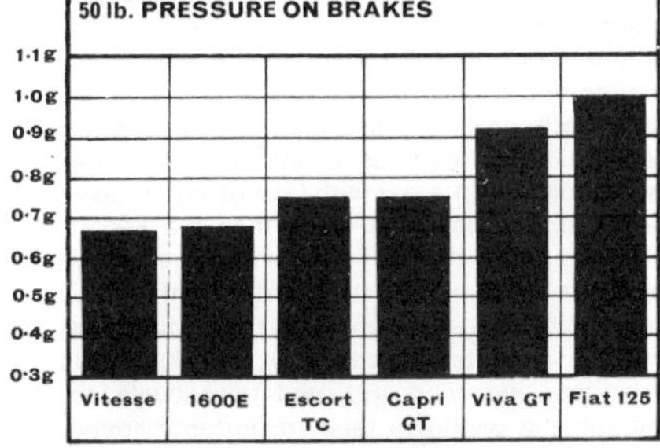

Comfort and controls

Best overall scorer in this category was the 1600E, confirming that its Executive tag is a worthwhile addition to what is basically a Cortina GT. Its nice-looking seats now slide on proper runners and are long enough to give good thigh support; the back rest is adjustable and grips well but could provide more lumbar support for sensitive backs. Its driving position was one of the best, too, with a well-placed steering wheel, controls, and good heel-and-toe pedal positions. All-round visibility produced another good score and the inertia-reel belts were comfortable to wear. Since most of our testing was on wet roads, we were forced to pay more attention than usual to washer/wiper switch accessibility and the wiper pattern itself; it was a slight stretch to the combined Cortina knob but at least you only need one movement for instant cleansing; the pattern is offset to the driver's side but the sweep finished short of the screen edge, effectively magnifying the thickness of the pillar blind spot. On ride, the Cortina was about the same as the Capri and Vitesse in the middle of the group—good on most surfaces but a bit choppy on some irregularities.

While the claim of a grand prix driving position in the Capri is a bit far-fetched, we all liked its low position with a well-placed wheel, and good pedal spacing. The cushion was a bit flat, but the seat adjustment is good. Its two-door styling allows the rear seat-belt mounting to be in the best place alongside the rear seat.

Visibility puts the Capri well down; you can't see the rear deck for reversing and the thick rear panels constitute blind spots at angled junctions. Its ride was rated the same as for the 1600E, but its rear movements are sharper although it feels generally better insulated from road shocks. Its wiper pattern was slightly better than the 1600E's and the floor washer button good; the fiddly wiper switch, though, is difficult to find without looking.

The Fiat's overall comfort was similar to that of the Capri, its good ride (the best in the group) and Cortina-type visibility making up for what it loses in the driving position, which we still feel is a bit short-leg/long-arm for taller people. The seats have good adjustment, but are a little short in the cushion for good thigh support. The pedals are perfectly placed for heel-and-toe changes. With a washer button on the floor and a two-purpose wiper stalk on the column—normal and intermittent—the

Continued on the next page

Vital statistics

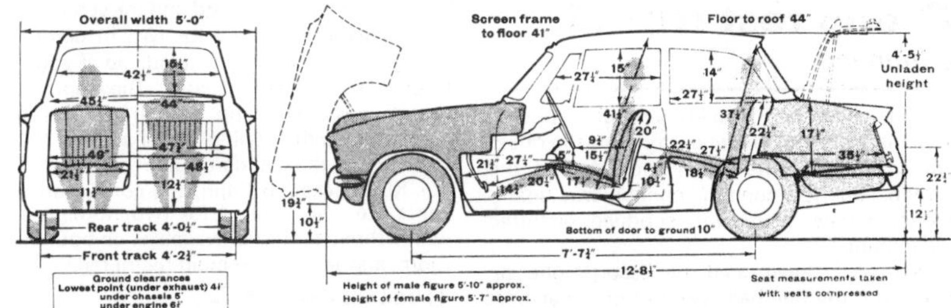

Triumph Vitesse 2-litre

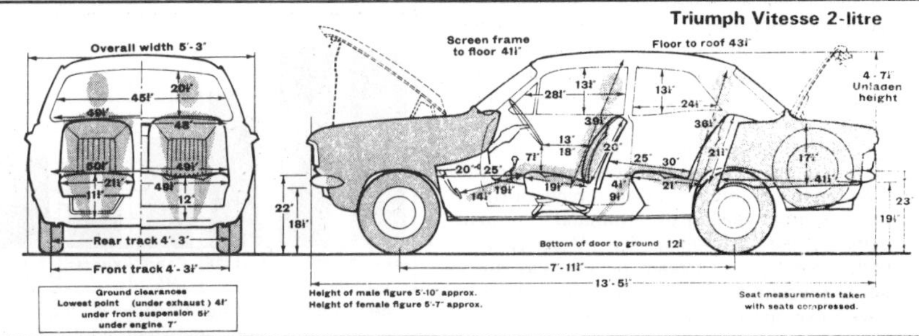

Vauxhall Viva GT

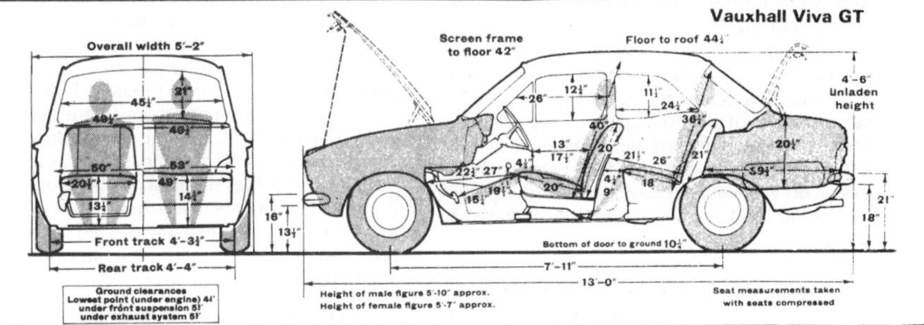

Ford Escort TC

Fiat 125

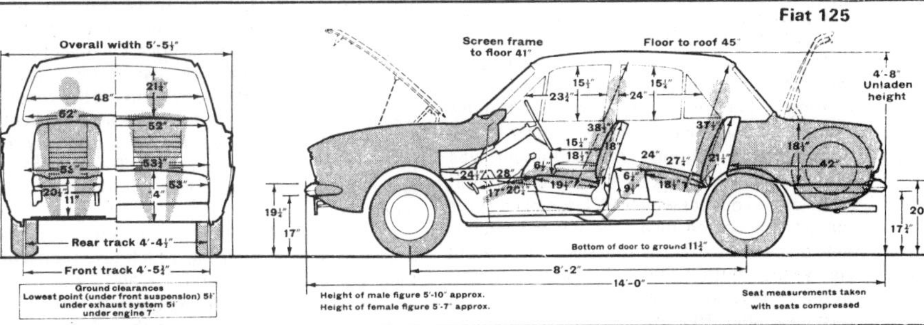

Ford Cortina 1600E

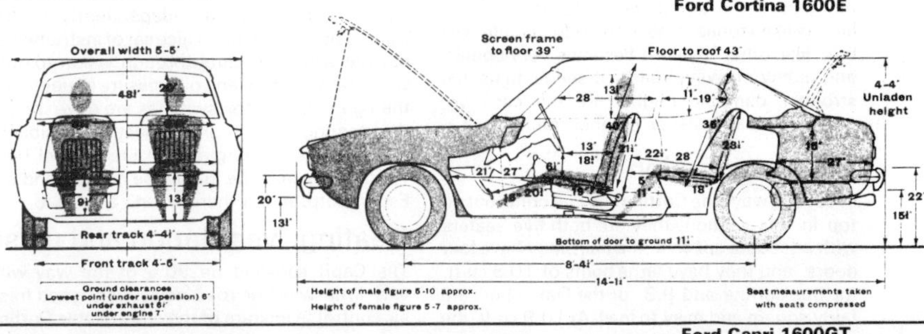

Ford Capri 1600GT

Group test No. 11
Continued

screen could be cleaned without removing your hands from the wheel though the offset wipers didn't quite reach the edge of the screen. On ride comfort, opinion was unanimous; it was quiet, with subdued, pitchfree movement.

We would have rated the Escort higher on comfort if the ride had been better; it can best be described as following the road surface faithfully, encouraging a sense of confidence in the car's considerable ability. But on anything but smooth roads it feels harsh and choppy and would probably be unpleasant on average Continental roads. The seat holds well and gives good support all round; it is set fairly high so that taller people have to place their knees carefully but this high position gives a great feeling of command; the seat belts earned top rating, too. Visibility was fairly good—the screen pillars and thick doors present a bit of a blind spot and you have to sit up to see the rear deck for reversing. The wiper pattern came up to the edge of the screen and the floor-mounted washer is good but the wiper switch is so badly placed down below the facia that inertia-reel belts are essential. Curiously, the Escort was the one car on which it was difficult to heel-and-toe; some said it was impossible, others persevered and succeeded by careful foot-placing. It is easier on unservoed Escorts where the pedal is heavier and usually depresses nearer the accelerator.

The Vitesse collected a selection of fair marks; the seat itself is quite comfortable with good support and good fixed angle, but none of us likes pedals offset to such an extent. Despite being the only one with an independent rear suspension, the Vitesse rated below the Fiat and on a par with the Capri and 1600E for ride comfort. It might feel better with a more rigid chassis construction. For its general controllability, the Vitesse earned high praise, but taller people complained of the shallow screen which hindered visibility on hilly corners; the wiper pattern was bad, too, though all-round visibility, provided the windows are clean, is excellent. The single wash/wipe button on the right was well placed.

Although the Viva looks very plush and impressive inside few of us much liked the seats, which were too short and too upright (non-adjustable), or the poor rearward adjustment which didn't allow you to get far enough back; the new, much lower, position of the steering wheel is a lot more comfortable though. Visibility was reckoned similar to that of the Capri although this judgment was probably adversely influenced by the very short wiper arc on our car which left large blind spots against the screen pillar; the wiper switch is well placed on the transmission console but the washer button under the heater controls on the facia is not, so the Viva's screen was the least cleanable. The ride is an odd combination; the suspension smothers small, sharp irregularities like stones, cracks and ridges quite well but, like other Vivas, it becomes very bouncy and pitchy on wavy roads. It seems to us that stronger damping might cure the car's uncomfortable turbulence on poor roads.

Accommodation

In several ways the Cortina and Fiat come out on top in this section; they are both five-seaters with adequate space in the rear, they have four doors, and they have large boots of 10.9 cu.ft. for the Cortina and 9.3 for the Fiat—both are fairly square and easy to load. At 10.9 cu.ft. the Viva is also well endowed and the semi-divided rear seats are shaped for two in comfort—though this could be a handicap if you have three children. At 7.8 cu.ft., the Capri's boot comes next, showing how much has been sacrificed to the styling—it is slightly smaller than that of the standard Escort at 8.0 cu.ft.; on our XLR-packaged Capri, the comfortable rear seats are also shaped for two, though a full width three-seater is available; on both, headroom and legroom remain restricted.

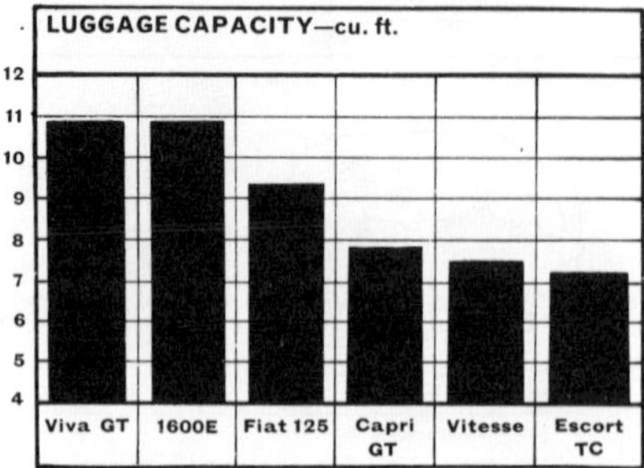

The interior of the Vitesse is the most cramped since two squashed adults is the maximum you can get in behind the front seats; its boot is a shade smaller than the Capri's at 7.5 cu.ft. too, but, as one advantage from its separate chassis, it doesn't need a high boot sill and thus has the most convenient loading height. The Escort TC's larger wheels dictate a revised spare wheel position which reduces its boot size to the smallest in the group; at 7.2 cu.ft. though, it is still larger than that of a BMC 1100, for instance, at 6.8 cu.ft; three people can just about fit into the back but two is really the comfortable limit.

Instruments and switches

On balance, the Fiat 125 came out on top of this section with a nice clear pair of dials containing all the required information, and well placed minor controls—a minimum of operations require releasing the steering wheel from your grip. Instruments on all the cars were good with the full specification which people like to see in this type of car; the Capri and Viva have the full panel visible through the steering wheel but they both suffer from poor switchgear, the Capri because its three cross-wise rocker switches are difficult to separate without looking, and because the light switch is over on the right. The layout on the Viva is much better than before but still untidy with switches scattered liberally on the tunnel, in the centre of the facia and around the dials. On the Vitesse the instruments are well placed for instant visibility and we particularly liked the big wiper/washer button. The lighting stalk arrangement is still bad, though, because it is too easy to turn the lights out inadvertently. The Fiat stalk, which does the same thing but in a different order, seems better.

On the 1600E, the instrument layout is scattered across the facia, which takes some of them out of easy checking line, and some of the switches are mean little things and unlabelled. With the Escort, it looks as though the facia layout was designed independently of the seating position; it has a nice set of instruments, but the wheel rim cuts through vision to both water temperature and oil pressure gauges, and the lights and wiper switches are down below the steering column and almost impossible to reach when wearing a non-inertia reel belt.

So on balance the Fiat came out top and the Escort bottom, but there wasn't a lot in it.

Heating, ventilation and noise

The Capri satisfied us 90% of the way with very little wind or road noise, very good fresh air control (a mixture of the controllable Cortina eyeballs and the cold up/hot down Escort versions), and it only needed a volume control on the roasting heater output to make it well nigh perfect. There is a volume control on the Fiat heater but the fresh air outlets only have a simple flap valve. On wind and road noise it was almost in the same league as the Capri.

Wind noise on the 1600E was also low provided that the door was shut properly (which was difficult to do), though there was some road noise. The excellent heating/ventilation, although like the Capri's, still needs a volume control. Fairly basic heater-ventilation equipment is used on the Escort and Viva but they both give the driver a cool head when required; the Viva, as usual, falls down on wind noise—on a par with Cortina and Vitesse—but it has less road noise than the Escort. The Vitesse is the only one without a proper cold air system, but its quarter-lights aren't too noisy when open and they lower the stuffiness quite successfully; its heater is adequate and wind and road noise levels fair, though the windows tend to flutter at speed.

In conclusion

Although the price tags are scattered around the £1,100 mark, the characters of these six cars differ widely from the raucous all-go Escort to the more luxurious and much more refined Fiat 125. The Escort is very much a road dicer with reasonably civilised amenities and space, but it is rather more tiring to drive than others in the group. High performance with more civilisation is offered by the Vitesse and Viva, the Viva being roomier and the Vitesse perhaps a little more fun and mechanically more refined with its smooth six-cylinder engine and overdrive.

As Ford's brand new car, the Capri was the newest design on test and does a lot to justify its clamorous launching. It is quiet and comfortable, reasonably quick, and very sure footed but it loses out on boot and back seat space.

Both the 1600E and Fiat 125 are slightly quicker than the Capri although not so well shaped for maximum speed; they score on being comfortable four-door five-seaters with large boots and, up to 50 m.p.h., they will out-accelerate an MGB. For the man who likes his sporty machinery to be capable of carrying a load of children or a couple of grandparents, they are by far the most practical.

The choice really lies between high performance and high comfort, or a compromise mixture in between. **M**

Previous group tests

May 18th Fiat 124, Ford Escort 1300 GT, Riley Kestrel, Saab V-4, Vauxhall Brabham Viva.
May 25th Austin Mini Mk. II, Reliant Rebel, Fiat 850, Hillman Imp, Renault 4.
June 1st MGB, MGC, Fiat 124S coupe, Triumph GT6, Triumph TR5, Reliant Scimitar.
June 8th Rover 2000, Triumph 2000, Sunbeam Rapier, Audi 90, Vauxhall Ventora, BMW 1600.
June 15th Austin 1800, Ford Corsair 2000, Ford Cortina GT, Hillman Hunter, Renault 16, Vauxhall Victor.
June 22nd Ford Escort 1100, Vauxhall Viva, Triumph Herald 1200, Volkswagen 1300, NSU 1000C, Austin 1100 Mk. 2.
September 7th Triumph TR5, Lotus Elan S4 coupé; Marcos 1600, Mercedes Benz 280SL, AC 428 Fastback.
October 19th Mercedes 220, BMW 2000TI Lux, Jaguar 420, NSU Ro80, Rover 3½-litre.
October 26th Singer Gazelle, Vauxhall Viva 1600SL, NSU 1200C, Triumph 1300, Toyota Corolla, Simca 1100 GLS.
December 28th BMC Mini Cooper, Fiat 850 Sport Coupé, Ford Escort GT, NSU 1200TT, Sunbeam Stiletto.

Next week

Personal choice—'Motor's' testers voice their individual opinions about these six cars.

Personal choice

*Five individual views of the six cars featured in last week's Group Test**

** Ford Escort TC, Cortina 1600E, Capri 1600 GT, Fiat 125, Triumph Vitesse, Vauxhall Viva GT*

I'M reluctant to single any one car out of this group, partly because each has considerable appeal of its own, partly because some are actually poles apart in character. So we are not really comparing like with like. For its combination of sheer performance and taut, responsive handling, the Escort Twin-Cam is out on its own as an exhilarating fun car. Moreover, this one was a lot less twitchy than the first we tried some time ago, and not quite so noisy either (though it was still the most raucous in the group) so I could live with it as family transport.

At the other extreme comes the Fiat 125, a car which leapt a couple of leagues in my estimation after this Group Test run. Its astonishingly smooth and quiet twin-cam engine—far more refined, I'm sure, than the last 125 we tried—its excellent performance, a superb gearchange and a roomy well appointed interior made it an unexpectedly attractive car which allies comfort, spaciousness and driver appeal to a degree that is rare in a car of this price. Only when I returned to my back-double commuting run was I reminded how heavy and sluggish the steering can be, and that it is very much better as a cross-country express than a town runabout.

The 1600E, on the other hand, fulfils both roles very well. Its comfortable seats, light responsive steering and excellent roadholding I'd rate more highly than the Fiat's but the engine is nothing like so smooth and effortless, particularly at high revs, so it lacks the mechanical refinement of the 125.

I find it very hard to choose between the other three cars. The Viva GT has a lusty performance, second only to the Escort TC's, but is rather noisy at speed; its cornering powers are very high yet it lacks the highly responsive handling that I like; and despite its lavish interior decor and equipment, it is not as comfortable as some of the others because the seat does not go back far enough. The lowered steering column and relocated switchgear have answered two of our earlier criticisms; with a few more changes, I believe the Viva GT could well emerge as a Group leader by combining in one car nearly all the individual virtues of the others.

The Capri and the Vitesse both rate highly as driver's dashabouts; each corners and handles well, each has a respectable performance, that of the Vitesse in particular being really quite aggressive. At the same time, it makes up in mechanical smoothness and peace what it lacks in wind noise suppression. The solid Capri is the other way about. Much as I enjoyed both these cars, though, they have less to offer as family transport than some of the others.

Roger Bell

I THOUGHT that this group was going to be made for me, all the quick saloons, mostly fast versions of standard family cars with enough performance and roadholding to keep me permanently satisfied; but somewhere something has changed. Either I have got further beyond the boy racer stage than I had realized and like too much refinement with performance, or the cars just aren't as good as I had remembered; possibly it is simply that the cost of cars has risen so much that my own bracket is really now around the £1,500 mark —which makes my choice rather difficult.

With tremendous performance and roadholding the Escort TC is a very strong contender and this one was a lot more stable in a straight line than the previous example; I have no qualms over reliability of the twin cam unit, so the car is capable of being used as an everyday transport, but everything about it is harsh. The low-speed ride is bad, even noisy, and the engine yowls tiringly when driven hard but perhaps a higher axle ratio would take the cruising speeds beyond the yowl period, which starts at 4,500 r.p.m. It is fun, but not for ever.

Progressing through from performance to comfort, the Viva comes next in that it is slightly more restful to drive; the ride is bouncy but not too harsh and the roadholding is probably as good as the Escort's thanks to its tyres—G800 Grand Prix. Performance is good but I don't really like its handling; it's a good car, but neither taut enough to be a real racer nor comfortable enough to be a civilized wolf.

If all schoolboys took as much notice of adverse reports as Triumph have done about the Vitesse, education would be a lot more painless a process for teachers. Its new rear end makes it a real sports saloon, its engine is extremely good, and it seems to have most of what I want; but certain details are too dated for me—cramped feeling inside, poor heating and ventilation, slightly unrigid feel. A near miss, though.

The 1600E had more character and a lot more luggage space than the Capri which is a lot more restful to drive, but I think the vogue for built-in understeer has been overdone, making it safe but uninspiring. The Cortina's good old recirculating ball steering has more feel than the Capri's rack. Of the two I'd prefer to own the 1600E, preferably a two-door one which would be slightly more rigid. A 1600E twin cam would be hard to beat. I'll learn to love the Capri later perhaps, when my staff 1600 arrives.

Which leaves the Fiat nicely placed at the comfort end of the scale; its ride is good, the roadholding not bad and its handling to my taste. You have to use full throttle to get good performance, since the second choke opens fairly suddenly, but the engine feels unburstable and will hum along smoothly forever. The 125 is comfortable and well made, doesn't feel at all like a boy racer but doesn't get left behind, either. So for all-purpose high mileage use I'd take the 125; but I'd like to have an Escort TC in the garage, just for fun.

Michael Bowler

TWO distinct groups in this bunch, Fords and others, all seeking to produce a car with equal appeal for a driver and his incumbrances. The difference lies in the balance between the two and it is extremely difficult to arrange five such entertaining cars in order of preference. As a driver's fun car the choice must be the Escort Twin-Cam which goes and handles almost like an Elan to show that the stork is not necessarily the kill-joy it used to be less enthusiastic wives may find it a bit fierce, a trifle noisy perhaps, cramped with only two doors to admit elderly relatives. Anyone would find the ride firm, minor controls a nightmare and the handling rather unforgiving beyond its very high limit, but if ever there was a reasonably practical fireball, this is it.

If you can learn to live with the understeer, the Fiat 125 shares many of the Twin-Cam's virtues including, of course, twin cams. This engine is magnificent with an insatiable appetite for revs and a smooth, well insulated flexibility to suit the well upholstered, spacious and fully glazed interior. Lack of adhesion could get a bit alarming in the wet but it's remarkably agile for a biggish car.

Like our elected leaders, Ford's engine department created a lot of noise and a mounting crisis with the BIP unit and left it to the rest of the organization to dig them out. They have done a much better job with the Capri than with the 1600E, booming resonance between 4,000 and 5,000 r.p.m. (about motorway speed) on the latter being reduced to a slight roughness around 3,000 r.p.m. on the Capri. Otherwise it's rather a question of whether you are prepared to sacrifice some of the 1600E's interior and boot space for the Capri's more sporting looks. They both have similarly first-class seats, driving position, heating, ventilating and that wonderful single rail gear shift. The Capri feels better put together and has lost most of the Cortina's body's wind noise. But I preferred the handling of the 1600E which has more weight and feel in the steering and for me just the right amount of oversteer.

Now that the Vitesse has truly exciting performance and a proper suspension system to match all it lacks is an up-to-date rigid structure to go with it. The engine delivers a smooth surge to the accompaniment of a contented howl from the exhaust when you accelerate and, although the ride is still a little bouncy on the rough stuff, it can be hurled round corners with scarcely a thought for what used to be the consequence of lifting-off. But inside it is still pretty horrid; the cramped hard-against-the-door driving position, rather sticky gearchange with its curious misaligned gate, crude heater controls, absence of ventilation and the wind noise from frameless windows flapping against their seals at speed remind you that the basic theme is beginning to show its age.

That little things mean a lot is exemplified by lowering the steering wheel of the Viva GT which has transformed its driving position, previously my chief objection to this car. Taken point by point I don't find it as satisfying as the Escort though it has many redeeming features in its comfort and equipment which makes it a rival worth considering. The engine has tremendous urge but is almost as noisy and lacks the Twin-Cam's smoothness. The handling, though still very chuckable, is not quite so precise, and understeer magnifies a feeling of lost motion in the steering. But as Vivas go the ride is reasonable and it is a lot of fun for the money and for me they could save on the matt black paint, multiple pipes and dummy hub-caps look. **Jim Tosen**

WE have found from our fast motoring in the Welsh outback that wet country roads tend to be much cleaner than city streets and to offer proportionately greater grip in consequence. But "Mud on Road" defined conditions for our last thrash because the tractors had been out in force—and if it wasn't low-mu clay it was ice. Adhesion assessments were therefore rather unreliable, depending more on the tyres fitted and the particular stretch of road traversed for each car; but for the same reason it was easier to judge handling since breakaway occurred at lower speeds.

A less doubtful advantage of the muddy roads was to bring out the importance of good wiper patterns and easily operated, quickly reached, wash-and-wipe controls. Clearing the screen on the Viva GT, for example, meant a fumble under the facia for the washer button plus a grope followed by a downward glance, to find the wiper switch on the central console. There were times when I wished that the designers of the layout were condemned to a week's Group Test in the wet sandwiched between Makinen and Hopkirk driving mudguardless cars. Yet the Vauxhall was the second quickest car in the group—on acceleration at least—had very good adhesion on its Grand Prix G800s and handled well in a rubbery understeery sort of way.

At the opposite end of the control layout scale was the Fiat 125 with its fingertip wiper stalk and floor button for the washers —the intermittent wiper action was also most useful in the dirty weather. Other virtues were the airy body, the commanding view of the road and the general quietness and refinement. It did a good job of telegraphing its slides, but in the conditions of our test the grip was as bad as the handling was good. The tail-wag on roundabouts was quite fun, but the unpredictable fits of the straight-ons at some corners were occasionally a little alarming.

If you like to sit *in* a car, rather than *on* it as you do in the Fiat, and to be surrounded by it, then you would like the Capri. I have only this high-waistedness against the car; it is a space-wasting styling compromise—but this is reasonable for the market sector at which it is aimed. Competent in everything it does, I found it rather uninspiring.

In many ways I preferred the Vitesse which has been transformed for the better by its new rear suspension. I liked its smooth, powerful engine, and its neutral way of getting round corners with a power-inducable gentle oversteer, but was less enthusiastic about its shallow screen and poor wiper pattern.

The 1600E was a good car, I thought, but why buy one for £1,097 when you can get the much faster Twin-Cam for only £98 more? Then you've got high-grade sports car handling and performance with room for the family and their luggage as well. Irritating though, that the most sporting car of the lot was the one in which it was most difficult to heel and toe. I'd want to bend the pedals about a bit, and fit a slightly less powerful servo to make it easier to pivot my foot on the brake pedal.

Anthony Curtis

FOUR of the cars in this group had only two doors and would therefore never qualify for a place on the Watkin driveway—four doors are essential with a large young family.

The Cortina 1600E looks right and is right. Tastefully trimmed and very well finished with comfortable seats and a good driving position. The car feels and behaves like a Cortina should; performance is brisk, the gearchange and brakes excellent and the handling and roadholding were such that it could be thrown around Welsh lanes like a Mini. With plenty of room for the family and their accoutrements it was just a pity that it was marred slightly by engine noise at the top end. What about some NVH treatment, Mr. Ford? But I would like one very much, noise or no noise—and for its size it was surprisingly economical.

If the Fiat 125 had better steering and roadholding it would have topped the bill, and I have no hesitation in voting its gearchange the best which, against Ford competition, is saying something. I liked the solid feel of the car, the excellent ride, and the engine felt unburstable right up to its astonishingly smooth 7,000 r.p.m.

It felt roomier and airier than the 1600E but did not have the former's good driving position and was definitely short in the leg for me. I wish all manufacturers would put the screen washer button on the floor with the wipers control on a stalk on the column —so simple really.

Much as I enjoyed playing at boy racer in the Escort TC I found it unpredictable in the wet and after a couple of unexpected heart stopping slides I decided that my family would be better off in the Capri. In many ways it had most of the ingredients I look for in a car of this type. Its worst failing was the harshness of its ride on poor roads.

The Capri was virtually unstickable. This must be the safest car of the year—even a fool would have a job to run out of road. The car felt beautifully balanced and sure footed and much as I liked the steering it gave little or no warning when the limit of adhesion was reached. Ford's NVH treatment has obviously paid off, the engine was quiet and remote compared with the 1600E but as a styling exercise it is both impressive and unfortunate—I don't see why rearwards vision and boot space should be sacrificed for the sake of appearance.

The Viva GT needs a little refining. The potential is there but the notchy gearchange was disappointing and although the lowered steering column is better, the driving position is still cramped. The switches on the centre console are an improvement on the tiddly affairs below the facia of earlier models and its roadholding was quite remarkable, and so were the brakes.

The gearchange on the Vitesse could have come from the same mould as that of the Viva but what impressed me most about this vintage machine was the way the improved rear suspension had transformed the back end. I never thought I would hang the tail out on a Vitesse and enjoy it. The engine is the best part of the car, smooth and full of torque which at full chat is a noise worth hearing. I found the narrow body limited elbow room considerably and why can't the people at Triumph's reposition the door handle—my right leg was quite sore after our two day excursion.

Barry Watkin

PHOTOS BY GORDON CHITTENDEN

FORD CORTINA 1600 GT

England's most popular car isn't really very English—but it is a rather nice car

SINCE THE ADVENT of the Maverick, Ford seems to be pushing the Cortina harder than ever. The two are certainly in direct competition with each other in this country, the basic Cortina undercutting Maverick's base price by about $140 and our test GT running about $110 more than last month's fully optioned test Maverick.

If you're determined to buy your economy car from a Ford dealer, there's a very clear choice between the two. If quietness, a smooth ride and fashionable styling are your bag, get the Maverick. If you prefer nimble handling, a little something extra in braking, a bit more entertaining driving and a larger trunk (the interior accommodations are quite close), the Cortina is your way to go. However, as we found after driving both the test GT and an automatic-transmission Cortina Deluxe (the basic $1960 model) neither of them offers as much performance for the price as the hottest similarly priced imported sedans.

The Cortina has received a greatly revised engine since our last GT test (October 1967). Displacement went from 1500 to 1599 cc and a new crossflow head—intake on one side, exhaust on the other—improved breathing. Power is thus up from 78 bhp @ 5200 rpm to 89.5 @ 5400—though the home version, without emission control, develops 93 bhp. Torque, however, is nearly unchanged at 98 lb-ft @ 3600 rpm; it was 97 @ 3600. The crossflow engine is somewhat noisier and rougher than the the old 1500, but its sound (a rather hashy sound due to its thin-walled exhaust headers) is consistent throughout the rev range normally used with no serious vibration periods until it gets up to 4700 rpm, a speed which will be encountered going up through the gears vigorously but not in American-style highway cruising.

The performance figures we obtained are slightly puzzling. The present GT and the earlier one both had difficulty getting off the line in the standing-start acceleration tests, being reluctant to spin their wheels, but though the times to speeds up to 60 mph are slightly less in the new car, it actually took longer to cover the ¼-mile than the earlier car. Comparing the present acceleration figures to other cars of similar size, weight and power, we find the present figures right in line

FORD CORTINA 1600 GT AT A GLANCE

Price as tested	$2408
Engine	inline 4-cyl, 1599 cc, 89.5 bhp
Curb weight, lb	2045
Top speed, mph	96
Acceleration, 0–¼ mi, sec	19.8
Average fuel consumption, mpg	22.0

Summary: pleasant sedan with attractive interior . . . willing engine, good gearbox . . . powerful but over-eager brakes . . . large trunk . . . good ventilation at road speed.

FORD CORTINA 1600 GT

so we must conclude that the 1967 car was an uncommonly good one. We might note too that the present test car had relatively few miles on its odometer—right at 1000—and might do a bit better with more break-in. But it *did* manage a substantially better top speed than the 1967!

The 4-cyl engine drives through a well synchronized gearbox whose ratios seem appropriate. The shift linkage is crisp and precise, but a little stiff on the test car; getting into reverse requires a bit of muscle as the lever has to be pushed down smartly to get into that gate.

A first impression in driving the Cortina is that the steering is heavy. Not as heavy as the Maverick's, however, and considerably quicker. Around town the Cortina's compact size and the quick steering make it nimble and out on the road its behavior is stable and safe if typical for a live-axle car. It is a car that can be thrown about enjoyably and safely; moderate understeer prevails unless the driver consciously chooses to drive it into a corner so hard that the tail comes out, and even when this happens the transition to oversteer is smooth and catchable. There is considerable body lean as in most small sedans.

The ride is on the choppy side, being rather sharply damped. With only the driver aboard there is plenty of suspension travel for vigorous driving on rough roads, but with four passengers it becomes easy to bottom the suspension. Only in braking, however, does the GT do anything really remarkable. Our panic-stop test from 80 mph brought out an outstanding deceleration rate (30 ft/sec/sec maximum, or 0.93g) and stopping distance (296 ft). But with the powerful braking comes rather dramatic behavior: the rear wheels are the ones that lock up, it is all too easy to get them locked because of an over-eager power assist, and the car wants to go sideways. The brakes' fade performance is good, no fade occurring until the fifth of our six ½-g stops from 60 mph, but there was a distinct pull to the left from the third stop on. In short, the Cortina GT just misses having an excellent set of brakes.

A British tradition has been upheld in the GT: a beautiful piece of wood, finished in clear lacquer, into which complete and highly readable instrumentation is set. The steering wheel is a bit high and the seatbacks are not adjustable for rake, but we've found much worse driver seating positions and the seats are fairly well contoured and padded. Ford has scored on the 3-point seat belts; they are simple, easy to hook up and easy to stow when not in use—there are even prongs for the inboard, female portion, something we hadn't seen on an import before.

The Cortina pioneered good dash-level ventilation in small sedans and its two dash nozzles still do as well as any others we can think of. Oddly enough, the rear quarter windows don't open at all, a striking cheapness. After spending considerable time in cars without front ventwings, however, we appreciated the Cortina's large and correctly pivoted ones in the front doors—we're not yet convinced that all cars can get along without them. A final comment on the glass areas: the wipers haven't been changed over from their right-hand-drive positions and there is a disturbing blind spot in front of and to the left of the driver. This can be corrected by repositioning the wiper blades, but then they won't park as they should at the bottom of the windshield.

Ford dealers have an air conditioning unit they'll install for $300, an increasingly important item and one not offered with dealer backing on all economy sedans. The other add-on option, also a dealer-installed one, is an AM radio. White-wall tires are standard on the GT, but radials can be ordered for a paltry $10 extra.

Cortina's appeal, like Opel's, seems to lie chiefly in the availability of dealers because of its tie-in with American Ford. Its GT interior is very attractive, and its brakes if not quite perfect at least do well at the feet of an intelligent driver. Otherwise it's very much like its competitors and it doesn't seem the bargain it used to be when compared, especially, to its Japanese rivals.

ROAD TEST
FORD CORTINA 1600 GT

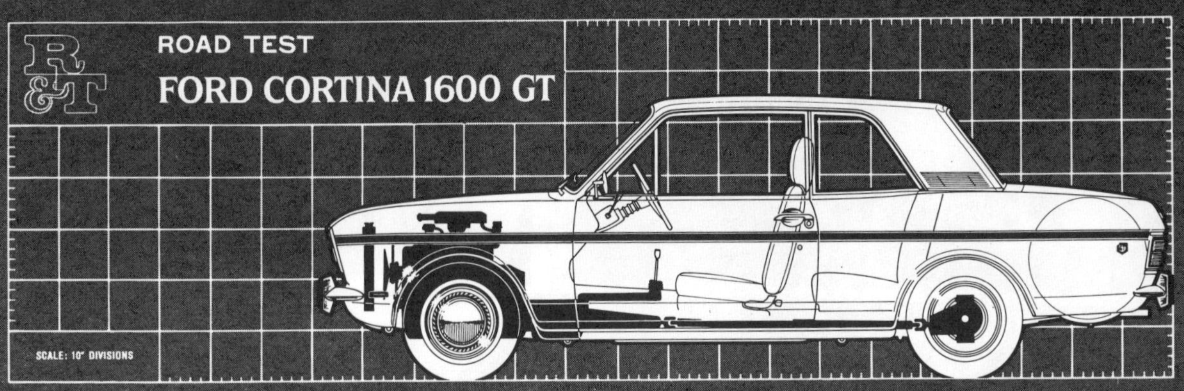

PRICE
Basic list..................$2408
As tested..................$2408

ENGINE
Type............4 cyl inline, ohv
Bore x stroke, mm.....81.0 x 77.6
Equivalent in......3.19 x 3.06
Displacement, cc/cu in...1599/97.5
Compression ratio..........9.0:1
Bhp @ rpm..........89.5 @ 5400
Equivalent mph..............95
Torque @ rpm, lb-ft....98 @ 3600
Equivalent mph..............63
Carburetion....one Weber 32 DFM
Type fuel required.....premium
Emission control......air injection

DRIVE TRAIN
Clutch diameter, in............7.5
Gear ratios: 4th (1.00)......3.90:1
3rd (1.40)...............5.45:1
2nd (2.01)...............7.84:1
1st (2.97)..............11.59:1
Final drive ratio............3.90:1

CHASSIS & BODY
Layout....front engine, rear drive
Body/frame............unit steel
Brake type: 9.6-in. disc front, 9.0 x 1.7-in. drum rear; vacuum assist
Swept area, sq in..........286
Wheels........steel disc, 13 x 4½
Tires.........Firestone F7 6.00-13
Steering type....recirculating ball
Overall ratio.............19.2:1
Turns, lock-to-lock..........4.5
Turning circle, ft...........30.0
Front suspension: MacPherson struts, lower lateral arms, coil springs, tube shocks, anti-roll bar
Rear suspension: live axle, leaf springs, tube shocks

EQUIPMENT
Standard: whitewall tires
Optional: AM radio ($65), air conditioning ($300), radial tires ($10)

ACCOMMODATION
Seating capacity, persons.......4
Seat width, front/rear.2 x 20.5/53.0
Head room, front/rear..39.0/35.5
Seat back adjustment, deg......0
Driver comfort rating (scale of 100):
Driver 69 in. tall............95
Driver 72 in. tall............70
Driver 75 in. tall............70

INSTRUMENTATION
Instruments: 110-mph speedo, 7000-rpm tach, 99,999.9 odo, oil press, water temp, ammeter, fuel level, clock
Warning lights: alternator, brake system, directional signals, high beam

MAINTENANCE
Engine oil capacity, qt.........4.2
Every 6000 mi: chg eng oil & filter, cln fuel filter, var. op'l chks
Every 12,000 mi: cln emission valve
Every 18,000 mi: chg air filter, torque rear spring bolts
Every 30,000 mi: pack front wheel brgs
Tire pressures, psi.........24/28
Warranty period, mo/mi.12/12,000

GENERAL
Curb weight, lb.............2045
Test weight.................2420
Weight distribution (with driver), front/rear, %...........55/45
Wheelbase, in...............98.0
Track, front/rear.....52.5/51.8
Overall length..............168.0
Width....................64.9
Height....................54.7
Ground clearance, in..........5.2
Overhang, front/rear...28.0/42.0
Usable trunk space, cu ft.....13.1
Fuel tank capacity, gal........12.0

CALCULATED DATA
Lb/hp (test wt)..............27.1
Mph/1000 rpm (4th gear).....17.5
Engine revs/mi (60 mph).....3430
Engine speed @ 70 mph.....4000
Piston travel, ft/mi..........1755
Cu ft/ton mi................79.8
R&T wear index...............60
R&T steering index..........1.35
Brake swept area sq in/ton....236

ROAD TEST RESULTS

ACCELERATION
Time to distance, sec:
0–100 ft...................4.3
0–250 ft...................7.2
0–500 ft..................10.8
0–750 ft..................13.8
0–1000 ft.................16.6
0–1320 ft (¼ mi)..........19.8
Speed at end of ¼ mi, mph....69
Time to speed, sec:
0–30 mph..................4.8
0–40 mph..................7.0
0–50 mph.................10.0
0–60 mph.................14.2
0–70 mph.................21.2
Passing exposure time, sec:
To pass car going 50 mph..7.3

FUEL CONSUMPTION
Normal driving, mpg.........22.0
Cruising range, mi............264

SPEEDS IN GEARS
4th gear (5480 rpm), mph......96
3rd (6000)....................74
2nd (6000)....................51
1st (6000)....................35

BRAKES
Panic stop from 80 mph:
Stopping distance, ft.......296
Deceleration, % g...........93
Control..................fair
Fade test: percent of increase in pedal effort required to maintain 50%-g deceleration rate in six stops from 60 mph..........29
Parking: hold 30% grade......yes
Overall brake rating........good

SPEEDOMETER ERROR
30 mph indicated......actual 29.8
40 mph.....................40.0
60 mph.....................60.8
80 mph.....................83.0

ACCELERATION & COASTING

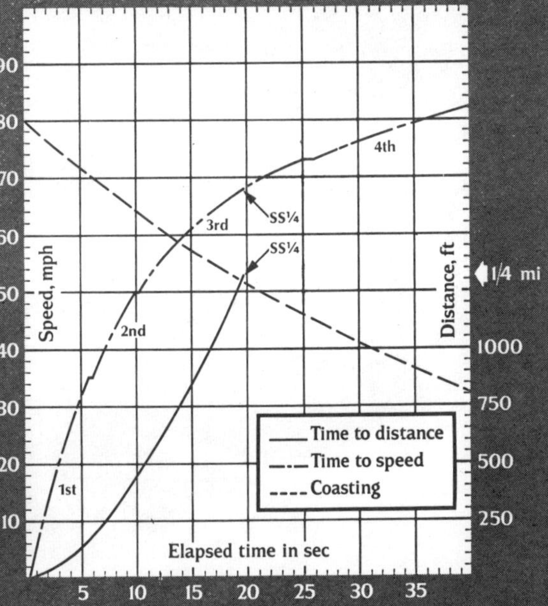

We evaluate the middle three and ask . . .

ARE THE GTs REAL?

SCW's technical staff take a searching look into the use of the revered term with the top three Australian manufacturers to find the answer.

IF insurance ratings are any criterion, far be it for us to say the Cortina GT doesn't earn its title. Surely its Baptism of Fire to be accepted for the *In* group could not be doubted. In black and white, the Cortina GT is a Group Four car — and we're not referring to FIA classification.

The same ratings set out by the Fire and Accident Underwriters' Association name the Cortina's next-of-kin, the Falcon GT, as a Group Three car. In terms of straight performance risk, the placid Cortina GT would scarcely rate past Group Two level while the tempestuous Falcon GT must immediately confine itself to Group Four.

Which all goes to prove that while the enthusiast regards the GT tag on a Cortina a mis-naming when considered alongside a true Ferrari, Aston type GT, the insurance companies don't. Tack GT, S or SL on and take the instant consequences. A normal Toyota Corolla will run 85 mph and a 20.0 sec quarter. Add SL, which gives you an extractor, tacho, matt boot stripe and a radio and you have 87 mph with 19.6 for the quarter. Meanwhile the insurance rating has descended from Group One to Four.

So a GT — whether it be just one step from a sedan hack or a full house Jaguar E — will cost big money to put and maintain on the road. Which brings us to the point of the exercise: do the current spate of sedan GTs deserve their titles and are they any competition for the outright sports car?

The frontal treatments of our three test cars bear striking resemblance. All have styling nose dip, all have a grille surround. The Hillman GT is the only one to feature square lights while the Cortina GT is the only one to have parker/flashers up beside the headlights.

If race successes are any guideline, the Cortina GT, with its history of race wins, deserves to be awarded the GT title over the two most recent acquisitions to the sporting sedan field, the Hunter GT and Brabham Torana. For three years up to 1965 Cortina GTs proved their salt under the pressures of Bathurst's annual 500. The current car, released after the Cortina GT500 Bathurst specials were banned from the Mount, is not the "race-bred" model, which can give one "that winning feeling". Many enthusiasts regard the Mk 2 Cortina GT as allying itself more with conformity. It has a definite stumpy, squat International look which provides greater frontal area and wind resistance than the "dart" shaped Mk 1 Cortinas.

The new body has also burdened the 1600 engine with extra weight which taxes performance. Ford's answer is the cross-flow head. This ups the bhp from 83.5 to 89, but leaves the Mk 2 down on the performance of a well-run-in Mk 1. The impeccable road manners and excellent gearbox are still there for the enthusiast to revel in.

Compared with the Hunter GT, the Cortina has a slightly wider front track and narrower rear track, which make it steer and balance more precisely. In all situations, the Cortina will remain quite predictable and only dire mismanagement

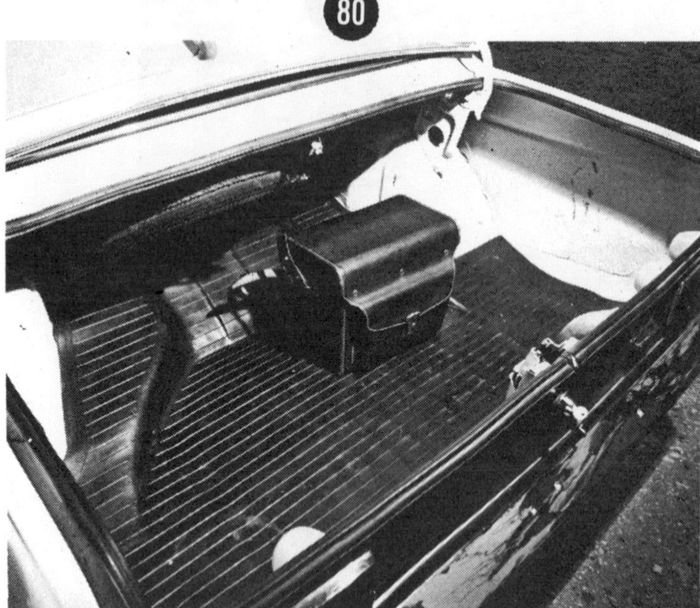

will see a tight position become an emergency.

The basic car has sufficient reserve in body strength, suspension and braking to withstand the enormous pressures placed on it by ace drivers such as Roger Clark who can perform quite superhuman acts of driving control. With insurance and registration thrown in, a new GT on the road costs over $2700 — but a true Cortina fan will always maintain he gets value for money.

The less enthusiastic will ask whether the slightly improved performance over a manual 1600, a gearshift console, tachometer, extra gauges, carpet, matt grille, metallic paint finish and ego-boosting GT badges are worth the extra coin.

The Cortina GT goes as close to being a true GT as any sedan will. It does not offer fire-breathing performance but it will respond to good driving. The steering, braking, gearchange and handling will all answer to fingertip finesse, which is a quality found infrequently in volume-produced sedans.

The Hunter GT still has overtones of the homely Hillman, a quality which it finds hard to shake. Whereas the Cortina has been improved from lessons learnt from racing, the Hunter is really a plushier, faster and better-equipped family car than the normal version. It is a tourer, not a Gran Turismo in its present form. The car itself has emerged as the lightest and most easily-handled Hillman for some time and not too difficult for women to drive. The steering column at last points in the same fore-aft plane as the car, which makes steering, gearchanging and braking easier than in its predecessors. While the steering and gearshift are lighter, they still lack a precise, sensitive feeling. The handling is no longer all understeer plough, synonymous with Hillmans. The stiffened suspension will allow the driver to indulge in some quite daring tail-out feats which in most cases are little more than just a neutral-oversteer-neutral flick. Slight understeer is still predominant with even tyre pressures, but this can be adjusted to individual requirements.

The tough, rugged image for the Hunter is emphasised by the London-to-Sydney effort. But the new GT— which is not so far divorced from the Marathon car (but for complete line welding replacing spot welds on all body joins and other trivia) — does not necessarily warrant the specific title because of this or any of its UK rally successes. Maybe it should be called Hunter LS.

For all its shortcomings in a claim for GT status the Hillman does offer excellent value-for-money. It is a very complete car with creature comforts down to headrests, full seat belts, heater-demister and reclining, well-shaped seats. The solid, five bearing crank, 1725 cc engine verges on the unburstable. The finish is good and undoubtedly production will remedy any recurring faults.

General Motors' token gesture at capturing a sporting image for the Torana range does not misuse the GT tag. The Brabham Torana is a match in handling for the Cortina and Hunter GT but lacks good torquey power. It is easy to drive the Brabham quite normally and still find the tacho running endlessly into the red for no appreciable turn of speed. The car is undergeared but if it were given a higher final drive the small engine would find life just as hard. The power to weight ratio dictates hard revving for some performance but short bearing life; or, slow-revving for little performance and higher bearing life. As the Brabham Torana is a "performance

The Cortina scores on boot size with 21 cu ft of space, while the Hillman has 18 and the Torana 10.9 cu ft. The Hillman is the only one with a boot light operating when boot is opened.

Cortina has the highest rear deck. Both Torana and Cortina have boot locks on the door, Hillman has it mounted on rear panel. Hillman is only one without wrap-around tail lights.

Engine-wise, the Cortina and Hillman share a 5-bearing crank while the Torana has a three-bearing unit. Cortina has cross-flow head and shares the same unit made of aluminium with the Hillman, although the Hillman is side-flow like the cast head of the Torana.

package", GM has taken the first route. The result is a car which would do well to have a 1500 or 1600 cc engine but in the meantime has to be content with small capacity and high rpm.

Where the Torana can establish true sporting performance is in its handling. The front suspension is quite conventional. The rear is by coil springs which are well-located with trailing arms.

The result is a sedan which will run rings around most small sports cars. The ride suffers from the quite soft rear coils which produce some fore-aft pitch. Although the pitch is no worse than others in the Torana's size class, the ride is not as comfortable as in the Cortina and Hunter.

The sedan GTs do not provide any foil to the high cost of sports car ownership. They rate right alongside the genuine sports car in the higher insurance bracket. This gives the sports car buff no qualms when he indulges in a no compromise sports car. The sedans cannot match the true GT or sports car for real driving response and performance. As a practical compromise, the GT sedans do fulfil their function admirably. Both the Cortina and Hunter can offer near-100 mph performance and comfortable, if not grand, touring with seating for four and, at a squeak, five.

Although the GT sedan cannot compete with the true GT sports car — and they do not really earn their GT title — they do fill a hole in the market. The Hunter GT is excellent value for money. It has copious amounts of the right gear but just misses the boat when it comes to performing the task of a GT. The Cortina GT goes closer to the GT mark in performance-handling but is still hampered by its obvious sedan connotations. The Brabham Torana does make fair pretence at being a with-it sports sedan. For the same price, the MG Midget, with all the disadvantages of a sports car, rates as better value for money. Then you don't have to rate the Torana as a GT — it bears the name Brabham.

Manufacturers have at last seen the light on bright color appeal. Drab colors are all very well for not showing dirt but when people buy cars that *are* a bit different, they expect them to look a bit different. When we gathered the three cars together for comparison they made a bright, pretty sight, if somewhat garish with the superfluity of GT striping. The comparison was done in the Christmas break while factories were on holidays. Instead of a collection of motley easy-clean colored factory road test cars, we had dealers' demos which were spotless and bright. The Holden came from C. V. Hollands of Rockdale while Peter Warren Ford, near Liverpool, supplied an immaculate red Cortina GT. The Hunter GT, in a lemon yellow slightly lighter than the Torana shade, was supplied from the manufacturer, Chrysler Australia.

The new Hunter GT sets a new standard in the sedan GT category, even rivalling the Fiat 125 for value. It has comparable straight-line performance with the Cortina's, has 1 ads more goodies and is $71 less. The Fiat 125, which is $372 more, has only a marginal lead in performance and offers few extras that the Hunter GT doesn't have, apart from the academic oneupmanship of twin overhead camshafts and other mechanical exotica.

As performance versions above cut-and-thrust ordinary hacks, the "GT" sedans fulfil their set functions. But for a true GT or sports car, you have to dig considerably deeper. #

The Torana scores with an alloy-spoked woodrim wheel but loses with no headlight flasher and has at least 80 miles less fast cruising on one tank of fuel. It is also alone with no full-flow ventilation while Hillman scores with cigar lighter, heater/demister and blower fan. Cortina is alone with no reversing light, no exterior rear mirror.

	CORTINA GT	HUNTER GT	BRABHAM TORANA
SPECIFICATIONS:			
Engine	4 in-line, 5 bearings	4 in-line, 5 bearings	4 in-line, 3 bearings
Bore and stroke	3.19 x 3.06 in.	81.5 x 82.5 mm	77.7 x 61 mm
Capacity	1599 cc	1725 cc	1159 cc
Compression ratio	9.6 to 1	9.2 to 1	9 to 1
Cylinder head	aluminium, cross-flow	aluminium, side-flow	cast, side-flow
Carburettors	1 2-bbl Weber	2 Zenith-Stromberg	2 Stromberg CD150
Rocker cover	painted steel	painted steel	painted steel
BHP at RPM.	89 at 5500	94 at 5200	79 at 5600
Torque (lb/ft) at RPM	102.7 at 4000	107 at 4000	NA
Transmission	4 sp, all syncro	4 sp, all syncro	4 sp, all syncro
Final drive	3.90 to 1	3.89 to 1	4.125 to 1
MPH per 1000 RPM	17.2	18.4	15.6
Body-frame constn	integral	integral	integral
Suspension, front	Macpherson strut, coils, anti-roll bar	Macpherson strut, coils, anti-roll bar	coils, wishbones, anti-roll bar
Suspension, rear	live axle, leaf springs, radius arms	live axle, leaf springs	live axle, coil springs, trailing arms
Shock absorbers	telescopic	telescopic	telescopic
Brakes	disc/drum	disc/drum	disc/drum
Power assistance	yes	yes	yes
Steering type	recirc ball	recirc ball	recirc ball
Turns l to l	4.25	4.0	3.2
Turning circle	35 ft	30 ft	30 ft 9 in.
Wheels	steel disc	steel disc	steel disc
Tyres	6.00-13 (c-ply standard)	6.00-13 (c-ply standard)	6.20-12 (c-ply standard)
Make on test car	Goodyear G800	Goodyear G800	Dunlop Guardian
Optional tyres	radial	radial	radial
Price	$2605	$2526	$2247
DIMENSIONS			
Length	14 ft 0 in.	14 ft 1.5 in.	13 ft 5.6 in.
Width	5 ft 5 in.	5 ft 3.5 in.	5 ft 3 in.
Height	4 ft 7 in.	4 ft 8 in.	4 ft 5 in.
Wheelbase	8 ft. 2 in.	8 ft 2.5 in.	7 ft 11 in.
Track, front	4 ft. 4.5 in.	4 ft 4 in.	4 ft 3 in.
Track, rear	4 ft 3 in.	4 ft 4 in.	4 ft 3 in.
Ground clearance	6 in.	6.75 in.	5 in.
Weight	18.5 cwt	20.0 cwt	15.5 cwt
Fuel tank	12 gals	10 gals	8 gals
PERFORMANCE			
Top speed	95 mph	98 mph	85 mph
Third gear	73 mph	75 mph	75 mph
Second gear	51 mph	49 mph	47 mph
First gear	35 mph	33 mph	28 mph
Acceleration 0-30	4.0 secs	3.4 secs	4.8 secs
0-40 mph	6.1 secs	5.7 secs	7.2 secs
0-50 mph	8.5 secs	8.3 secs	11.3 secs
0-60 mph	12.9 secs	12.4 secs	16.6 secs
0-70 mph	16.5 secs	16.7 secs	21.5 secs
40-60 mph (top)	7.0 secs	8.7 secs	12.4 secs
50-70 mph (top)	8.7 secs	10.7 secs	14.7 secs
Standing ¼ mile	18.3 secs	18.5 secs	19.9 secs
Fuel consumption	24 mpg	26 mpg	22 mpg
EQUIPMENT			
Heater demister	no	yes	no
Blower fan	no	yes, 2 speed	no
Seat belts	lap	lap and sash	lap and sash
Windscreen wipers	1 speed	1 speed	speed
Anti glare arms	yes	yes	yes
Windscreen washers	manual	power	power
Cigarette lighter	no	yes	no
Exterior rear vision mirror	no	yes	yes
Gauges	fuel, oil, temp, amps	fuel, oil, temp, amps	fuel, oil, temp, amps
Tachometer	yes	yes	yes
Headlight flasher	yes	yes	no
Fuel tank range (fast cruising)	280 miles	260 miles	180 miles
Reversing lights	no	yes	yes
Through flow ventilation	yes	yes	no
Front seat squab support	poor	fair	poor
"GT" gear	blackened grille, rear flank GT badges	blackened grille, black boot flash, GT sill stripes, GT badges petrol cap and grille, simulated alloy spoked steering wheel	black bonnet stripe, black boot flash, Brabham badges on front flanks, woodrim steering wheel and gear knob.

IT'S POINTLESS THESE days to carp about the abuse by manufacturers big and small of the GT name. At present you can choose in Britain from something like 41 cars with GT somewhere in their title, made by 13 manufacturers. Out of this mass only a handful really qualify from the purist's point of view.

So in the last few years GT has come to signify a saloon with performance improvements covering engine and suspension, topped off with a few detail alterations in styling to ram home the point that the car is a cut above the others. For the maker, the whole concept is a godsend, enabling yet another 'new' model to be wrung out of the same basic body shell/engine/ running gear set-up. In most cases the higher price accurately reflects the value of the extras fitted. In a few, a buyer might make a useful saving if he went for the standard saloon and paid a reputable tuning outfit to uprate the car to the loftier level of performance.

This objection certainly does not apply to the Hillman GT and Ford Cortina GT examined here. The Ford needs no introduction; or it shouldn't, having been around longer than most of its rivals yet still topping the sales graphs against immediate competitors. The Hillman, on the other hand, was launched a month ago (with the 'Hunter' bit of the name deleted because Minx-style economy features don't justify it) in time to catch the Paris Show without losing whatever impact it might make on arrival at Earls Court a fortnight later. Its success in the market-place remains to be seen.

Unlike a Ford, no Rootes car is bought today for its sporting connotations. The open two-seater range has gone and with it the Ford V8-engined Tiger. The Rapier is something of a disappointment to drive—even in H120 form with Holbayised engine—and the patchy styling is not enough to attract the discerning on its own. No, Rootes from a pinnacle of competition success in the 1950s and a good enough run more recently have slumped to the point where the competition department has been shut down altogether.

But what of the London–Sydney Marathon? A storybook win, admittedly, and one that surprised Rootes as much as anyone else. Yet sales didn't take the anticipated leap upward. Nearly a year later, comes the model that would have gained most from that success. Unfortunately, the Marathon has already begun to recede into the mists of time and with it goes the last vestige of that all-important image. The Hillman GT is left to compete for customers strictly on its merits.

The Hillman and the Cortina are remarkably similar in specification as well as in appearance. And, as it turns out, in performance. There is just rather more to choose when it comes to price. A four-door Cortina GT is a mere £49 more than the Hillman but the extra cost of the Ford includes delivery to dealers on the mainland of Britain so the true difference when you come to write the cheque could well be lower, especially as Rootes are charging extra for a de luxe trim package.

STYLE AND ENGINEERING Considering the influence exercised over the British company by Ford in Detroit, where dignity and reticence don't exactly reign supreme, the Cortina GT comes out looking surprisingly innocuous. Devoid of speed stripes, large areas of matt black paint or ornate wheel trim, it appears from a distance very much like every other Cortina; you have to find another £100-plus to get the decorated exterior of the 1600E. The GT comes with two- or four-door body shell, the latter costing another £26 and well worth it.

The Hillman is offered with four doors only. It has been worked over on the outside with all the regulation Motor Industry Issue bits for GT-ising. The front grille, carrying rectangular headlights, has more black paint than chrome. Double stripes in a contrasting colour stretch along either side of the body at waist height. The wheels are black and silver Rostyle and to top it off small GT motifs are fastened to the rear quarter panels. Funnily enough, the overall effect is quite pleasing. It also helps the Hillman to bear a passing resemblance to the Hunter in which Cowan, Coyle and Malkin won the Marathon.

Rootes fit a leathercloth padded rim steering wheel with perforated alloy spokes where Ford stick to unadorned plastic.

Beneath the bonnets the overall similarity is maintained. Bore/ stroke ratios are nearly square in either case, the crossflow Cortina in fact being slightly over- and the non-crossflow Hillman less than a millimetre under-square. The Hillman benefits from an alloy head, while the Ford has the combustion chamber-in-piston arrangement that now extends most of the way through the Dagenham range. Both engines produce peak power at under 5500rpm yet are happy to run well above that figure, as indeed they should in this day and age. Both tachometers are red-lined at 6000rpm. The Hillman's power unit is in the same state of tune as supplied for the Rapier, with modified cams, valves and exhaust, breathing through twin variable-choke Strombergs, and altogether much milder than the lumpy 900rpm tickover suggests. For this application Rootes have put in a three-row radiator.

The Ford has a single compound-choke Weber and despite being down on cubic capacity and lacking the theoretical benefits of an alloy head is exactly equal on power.

Macpherson struts are used for the front suspension on both cars, as are live rear axles and semi-elliptic leaf springs. Ford, however, add radius arms to ensure accurate axle location. Rootes don't bother with this refinement and to judge by the handling and absence of wheelspin or axle tramp under full acceleration are justified in the omission. The only differences from the normal Hunter set-up—apart from wheels—are in details of the front end, where the firmer spring and damper settings of the Sunbeam Rapier are used, plus beefier stub axles from the Rapier H120 to cope with the heftier loadings imposed by wide-rimmed wheels. The anti-roll bar is also of larger diameter.

Gear ratios are reasonably well spaced in both cars. Overdrive, controlled by a switch on the steering column, is a £55 extra on the Hillman (not available for the Cortina) and is necessary to keep the Rootes car on a competitive performance footing. With it as part of the package comes a lower final drive ratio—3.89 to 1 against 3.7—and the effect of this is to make the Hillman a lower geared and therefore livelier car yet still with a long, loping top of 3.12 to 1 compared with the Ford's 3.9.

USE OF SPACE The similarity that crops up so often elsewhere between them is repeated in overall dimensions. They're equal in length, though the Hillman is an inch higher and one and a half inches narrower, and are half an inch apart in wheelbase, the Ford being the shorter.

This near-twinning is there again in the passenger compartment. The rear seats take two adults with some space to spare, at least as far as

The seats of the Hillman are thin, hard and unyielding with rather restrictive fixed headrests (top), whereas the Ford's optional reclining seats are the best they've done yet

▶ width is concerned, but a third back seat passenger is going to find things pretty tight. Particularly in cars like these, with pretensions to cornering ability, a single occupant of the back seat gets thrown around because of the absence of any lateral location.

Headroom is adequate by current standards. By the same measure there is sufficient legroom, although this means that, unless front seat occupants consent to sitting well forward, knees are pressed firmly into the squabs.

No one should find anything wrong with the front seats in either car as far as spaciousness is concerned. The transmission tunnels are reasonably compact. In the standard form the Hillman lacks a tray of any kind atop the hump although the near-bottom-of-the-range Minx De Luxe has one. The question doesn't arise on the Cortina because of the central handbrake.

Luggage space is similarly ample. In both cars, the spare wheel shares the boot with the suitcases.

COMFORT AND SAFETY

There's little to choose between the two in ride quality. Both are set up a shade firmer than the standard saloons from which they are derived but no one could accuse them of even approaching the point at which real harshness begins.

Under extreme stress, as in a sharp swerve around a tight corner, the Cortina exhibits more body roll than might be thought seemly in the circumstances. The Hillman is better under these conditions, and both stay decently near level during normal cornering.

The Cortina's driving position feels a shade old fashioned after the Hillman. There still seems to be a little of the 'control comes from a big wheel in your lap' outlook at Ford, though the Escort is an honourable exception. The Hillman puts the wheel lower down and farther away, and the seat higher, to give a position that is both modern and comfortable.

The pedals in Ford and Hillman alike are properly spaced, without noticeable offset either way. The Ford has a spring-loaded treadle on the end of the pedal arm which results in the driver's foot acting on a wide plate all the time. Clutches are medium in pressure required and have a sensible amount of active travel in taking up the drive.

Ford at last appear to be growing out of handbrakes under the dash, putting the GT's between the seats. Rootes prefer a location ▶

	FORD CORTINA GT	HILLMAN GT
PRICES	Cortina GT prices start at £960 for a 2-door, £85 more than the stock 1600cc Super. The 4-door model as tested is £986, slotting neatly in between the 4-door Super and the GT-plus-luxury-trim 1600E. Reclining seats bring the true price of the 4-door GT up to £1,012	By trimming the specification Rootes have kept the Hillman GT in standard form £29 below the Hunter at £962, where it is still £111 more than the Minx De Luxe. Reclining seats put another £16 on to make the GT price £978 and the highly desirable overdrive brings it up to £1,034
ACCELERATION from standstill in seconds	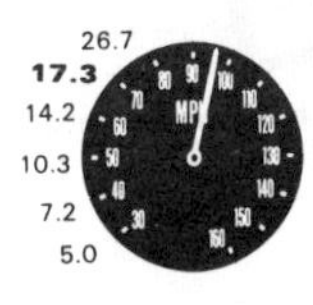 26.7, 17.3, 14.2, 10.3, 7.2, 5.0	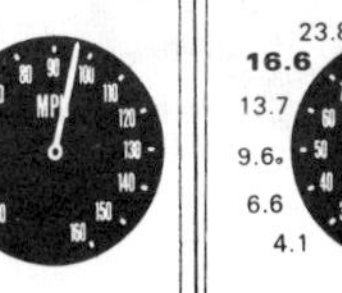 23.8, 16.6, 13.7, 9.6, 6.6, 4.1
FUEL	23 mpg overall ★★★★ 28mpg driven carefully 230–280 miles range 10gallons capacity	22 mpg overall ★★★★ 27mpg driven carefully 220–270 miles range 10gallons capacity
SPEEDS IN GEARS	36, 54, 78, 95 top speed	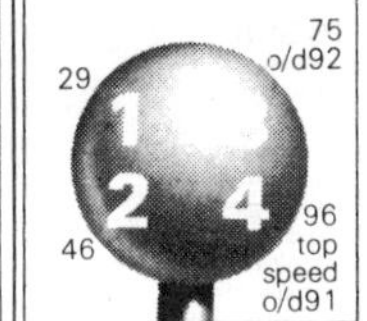 29, 46, 75 o/d92, 96 top speed o/d91
HANDLING	Roadholding good on fairly wide G800s, although heavy steering prevents full advantage being taken of stable handling. Strong understeer changes to oversteer only under extreme conditions and/or provocation. Basic characteristics unchanged in wet	Roadholding good on India Autoband radials on broad rims. Manageable understeer maintained until eventually overruled by progressive breakaway at rear, easily held on quite light and responsive steering. In wet, rear becomes rather sensitive
LUGGAGE CAPACITY number of 10in diameter spheres accommodated in boot	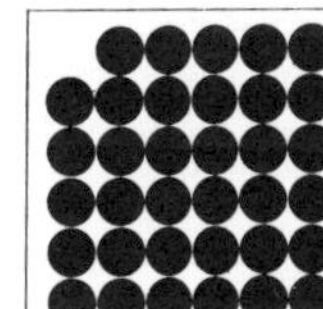 	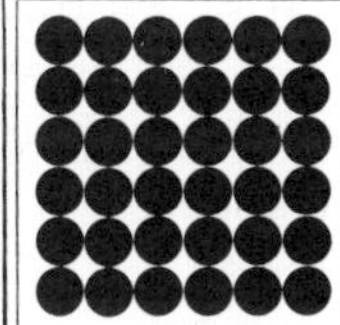

BRAKES RESPONSE in normal use. Deceleration (percent g) vs pedal load (lb) A = Hillman GT B = Ford Cortina GT

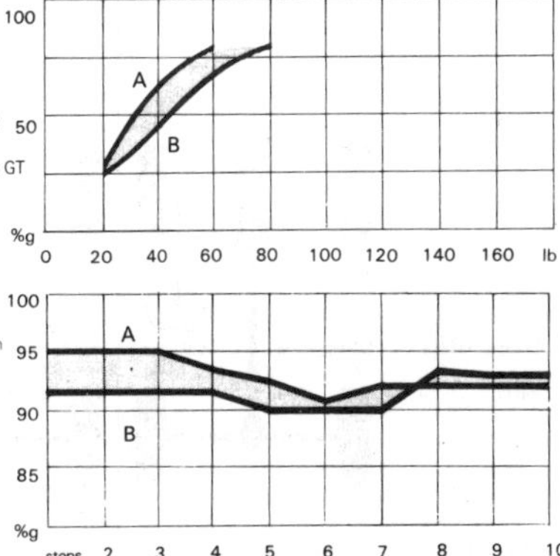

FADE peak deceleration achieved in 10 crash stops from 60mph at one minute intervals

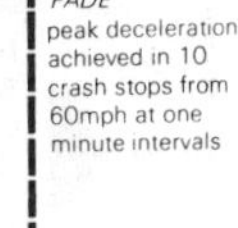

	FORD CORTINA GT	HILLMAN GT
DIMENSIONS	inches	inches
wheelbase	98	98.5
front track	52.5	52.5
rear track	51	52.5
length	168	168
width	64.9	63.5
height	55	56
ground clearance	5	6.75
front headroom	38	38
front legroom	24/28	22/30
rear headroom	37	38
rear legroom	24/27	22/29
ENGINE		
Material	iron/iron	iron/alloy
bearings	5	5
cooling	water	water
valve gear	pushrod ohv	pushrod ohv
carburettors	1 compound choke Weber	2 variable choke Stromberg
capacity cc	1598	1725
bore mm	81	81.5
stroke mm	77.6	82.55
compression to 1	9.2	9.2
net power bhp	88	88
rpm	5400	5200
net torque lb ft	96	100
rpm	3600	4000
TRANSMISSION		
control	floor lever	floor lever
synchromesh	1-2-3-4	1-2-3-4
ratios 1 1	11.59	12.41 std 13.04 o/d
2	7.84	7.92 8.32
3	5.45	5.15 5.41
4	3.90	3.70 3.89
		o/d 3rd 4.35
		o/d 4th 3.12
final drive ratio	3.9	3.7 (3.89 with o/d)
tyre size	165/13	165/13
rim size	4.5	5.0
SUSPENSION		
front	Macpherson strut, coil springs, telescopic dampers, anti-roll bar,	Macpherson strut, coil springs, telescopic dampers, anti-roll bar,
rear	live axle, semi-elliptic leaf springs, twin radius arms, telescopic dampers	live axle, semi-elliptic leaf springs, telescopic dampers
LUBRICANT		
engine oil type SAE	10W-40	10W-40
sump, pints	7.2	7.5
change, miles	6000	5000
other lube points	none	none
lube intervals	—	—

	FORD CORTINA GT	HILLMAN GT
AIR	24psi; recirculating ball	24psi; recirculating ball
BRAKES	disc 9.6	disc 9.6
STEERING	30ft turning circle; 4.5 turns lock to lock	34ft 2in turning circle; 3.75 turns lock to lock
AIR	28psi	24psi
BRAKES	drum 9in	drum 9in
WEIGHT	1991lb	2111lb (2131 with o/d)

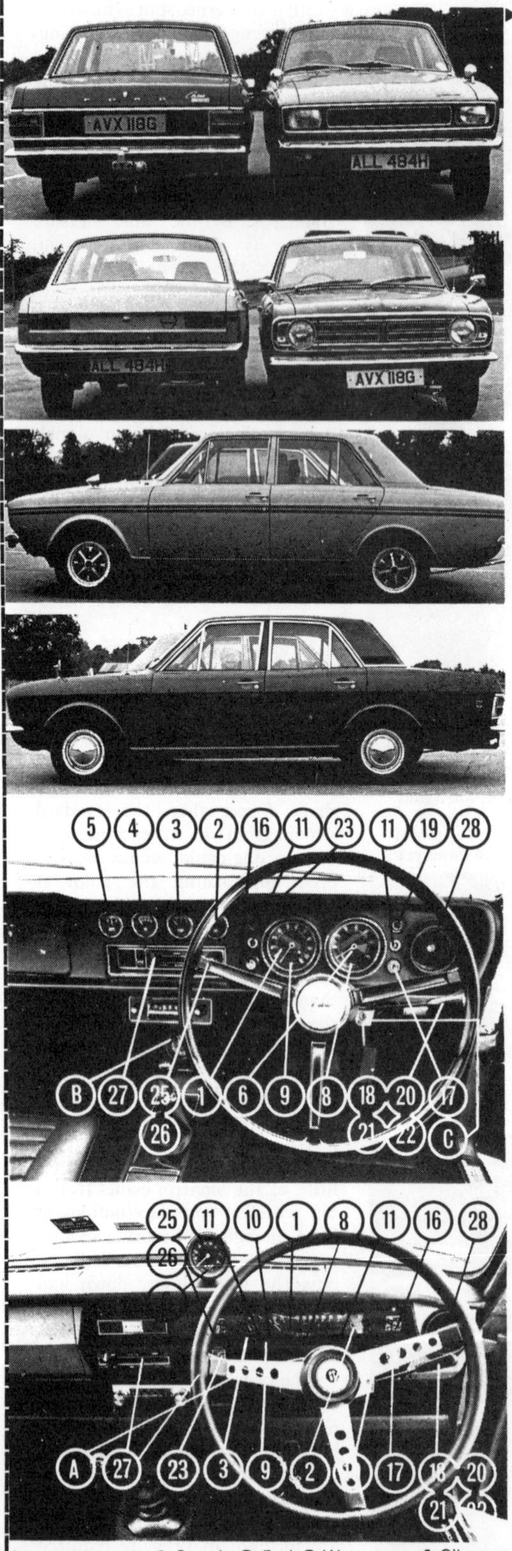

Instruments: **1** Speedo **2** Fuel **3** Water temp **4** Oil press **5** Amps **6** Tacho **7** Oil temp. **Warnings:** **8** Ignition **9** Main beam **10** Oil press **11** Indicators **12** Water temp **13** Sidelights **14** Handbrake **15** Fuel low. **Controls:** **16** Choke **17** Ignition/start **18** Indicators **19** Lights **20** Dip **21** Flash **22** Horn **23** Panel lights **24** Parking light. **25** Wipers **26** Washer **27** Heater **28** Face level vent. **Special Items:** **A** Overdrive **B** Clock **C** Towing warning light.

▶tucked down between seat and door.

In the trim department Ford stick to straightforward, functional looking leather-like PVC and carpet. The former material abounds in the Hillman, being used for everything except the floor covering. Here Rootes have forgotten the price and nature of their GT and put down cheap but more useful rubber mats.

The Cortina scores again on seating. The front ones are comfortable with enough firmness to prevent fatigue setting in on a hard day's drive. They give adequate lateral location and longitudinal adjustment. The Hillman's, complete with built-in neck restraints (of which more anon), seem all right until the car gets moving. Then it becomes plain that they are almost completely lacking in any sort of sideways location for the occupant, making safety harness a necessity for all but the most leisurely motoring whether you like belts or not.

There is no proper adjustment for the backrests on either of these near-£1000 GT saloons. Reclining front seats are £26 extra for the Cortina, £16 more on the Hillman. Illogically, the Rootes recliners have normal height backs.

Both cars have fixed quarterlights, eliminating a source of both extra ventilation and unauthorised entry but also lowering wind noise.

You can see all four corners of the Cortina without stirring far from a normal driving position. In the Hillman the neck restraints—which provide an intermittent cranial vibro-massage if used on the move—are a valuable safety item in their primary purpose but a nuisance for the way they obstruct the view when reversing or glancing over one's shoulder. In the rear view mirror they are nearly, but not quite, out of sight. These seats were forced on Rootes by the US safety regulations, were tried and allegedly approved of on the Marathon and now crop up as standard on the GT. Rootes clearly expect a few objections from the paying public, or from regular back seat passengers who dislike the restricted view presented to them, and are offering ordinary low-back seats as a same-price alternative. They are not available with the de luxe trim pack.

On balance, we think the neck restraints are worth having. Assault from the rear is one sort of accident you are unlikely to be in a position to prevent and the resultant injuries to the top of the spinal column as your head snaps back under impact can prove long lasting—like fatal.

On the Cortina the wiper pivots are placed far over to the right so that all of the driver's normal range of vision is swept clear. A more nearly central location is favoured by Rootes, giving the same result.

Noise levels have been kept generally low in both models. As elsewhere, the Hillman leads in most respects by a small margin. The Ford's engine coarsens to considerable roughness as it nears maximum revs and the effects of this are not only audible, for at 80mph engine-excited vibrations blur the mirror. Wind noise and road rumble are also more pronounced on the Ford, though the Hillman is not exactly silent at high speed. Neither car has a raspy enough exhaust note to indicate that it is anything other than standard.

Ford might have popularised the idea of through-flow ventilation—'Aeroflow' in Ford parlance—but other manufacturers didn't need all that long to catch up. The Rootes system is practically identical, even down to the outlet grilles in the rear quarter, though Rootes prefer an almost vertical location for them against the horizontal alignment on the Cortina. Either way, the result is the same with the systems doing a pretty thorough job of heating and a fair one of ventilating.

Padded ledges surmount the entire length of the facia panel on Hillman and Cortina alike but only in the Ford—now to US Federal safety specifications in this respect—is the padding continued round beneath the panel to protect the knees. Both have non-lockable compartments facing the passenger, plus parcel trays beneath. Only the Hillman's extends the full width of the compartment. In the Cortina it ends at the central console.

Having lifted the engine out of the Rapier, Rootes appear to have had a bout of amnesia and left that car's relatively comprehensive set of instruments behind. The entire Rapier dash would fit straight into the GT—at the production stage, that is, for owners would find the switch a major undertaking. The GT's special American-made rev counter, which can be turned around a vertical axis for adjustments, is an obvious and not altogether successful afterthought. The speedo tries to combine a horizontal strip dial with a needle moving around a central pivot. The result is that accuracy is not this instrument's forté. The only other gauges cover fuel level and water temperature.

The Ford scores heavily here, although the earlier layout with minor dials set into the padded lip has been abandoned on safety grounds. Speedo and rev counter are circular, unadorned, with white figures on a black ground. Then, ranged away to the centre of the dash, are fuel level, water temperature, oil pressure and ammeter, the latter just about due for replacement by the more informative voltmeter already in use elsewhere in the Ford range. Beneath this lot is a Keinzle clock.

PERFORMANCE, HANDLING, BRAKES Possibly because of a slightly more audible and decidedly deeper exhaust note, perhaps because all that stuff Ford used to put out about Total Performance really does penetrate the subconscious, everyone starts out expecting the Cortina to have the legs of the overdrive-equipped Hunter. In practice it's the other way round, though not to a truly appreciable extent.

All the way up the speed scale the Hillman leads, despite being heavier by a hundredweight. At first it is only fractions, for less than a second covers both cars up to 70mph. But from then on the Hillman surges away, scoring not so much on an excess of power—and certainly not through markedly better aerodynamics—but by having twice as many gear ratios to play with once you're out of second. To achieve this advantage the driver has to display some dexterity in juggling overdrive and gearbox. The Hunter is one of those comparatively rare cars on which overdrive third is actually of any use, being substantially lower than direct fourth. Thus as one reaches peak revs in third the next step is to flick up into overdrive on the steering-column-mounted stalk. Then comes direct fourth, knocking the overdrive out of engagement at the same time, and the engine is still biting in the fattest part of the power curve before it's time to go up into o/d top, which is more like sixth gear. At this point the overall ratio is getting a bit much for the engine, even with the lowish final drive gearing, and things begin to tail off as the maximum speed comes up.

The Ford meanwhile has a straightforward selection of four gears, quite nicely spaced with a usably high second gear. The Hillman has this feature too, but to such an extent that a sticker on the windscreen warns against trying to get away from rest on it (rightly, as it turns out).

With an engine 127cc down on the o/d Hillman's and with two fewer gears, the Cortina is really doing an excellent job in staying so close to its rival. The engine feels decidedly lustier, especially in the all-important middle part of the rev range. For main road overtaking between 40 and 70 it manages at least as well as the Hillman with a lot less gearchanging.

The Hillman is more refined in the exhaust and mechanical commotion department, and quieter, too, where wind noise is concerned. In the Ford at 90mph the roar of air spilling around the windscreen pillars is enough to drown out the other disturbance sources. It almost prevents conversation as well.

In matters of handling both cars conform to the best English traditions of understeer, and neither has enough power to let you achieve much with the throttle. In fact a snap opening or closing of it has little effect either way unless the car is already teetering over the ragged edge of adhesion. All this works as something of a built-in safety valve in that the more lock that is wound on to counteract the understeer the more speed is scrubbed off. By trying really hard the back can be made to go eventually, but both cars break away in progressive, easily manageable fashion. Of the two, the Hillman feels distinctly more controllable due to the fact that it is blessed with much better steering. The Cortina has a rather high-set wheel and the steering gets progressively heavier as the wheel is turned, suggesting excessive castor action. However, this is primarily a defect when seen in relation to the Hillman for the Ford's steering is not all that heavy as medium-sized saloons go.

Sheer cornering power is at a high enough level in both cars, the Hillman just having the edge perhaps although differences here are academic. Both cars come with radial tyres, India Autoband on the Hillman and squeal-prone Goodyear G800 for the Ford. The Hillman has an extra half-inch of rim width. The Cortina, incidentally, responds well to a few more pounds in the front tyres.

Medium saloons like these, even in improved performance form, are borderline cases for the use of a servo to supply power-assistance to the brakes. Ford don't trouble to fit one and a driver would need to be extraordinarily lacking in leg muscle to complain about its absence. The action is smoothly progressive right up to the point of rear wheel locking. This comes in early enough to justify the sort of pressure limiting valve in the back brake line that is common on front wheel drive cars. Weight transfer is the basic cause, of course, and it afflicts the Hillman in almost as marked a fashion. The Rootes car, though, has a servo, giving just sufficient boost to the brakes to take the edge off any incipient hardness without making them unduly sensitive.

On a 1 in 3 hill both cars would hold on the handbrake alone, the Ford's barely gripping on this incline, however, and not imparting overmuch confidence to the driver. Again, both could get away from standstill but the lower-geared Hillman managed to do so with a great deal more verve.

Taken all round, handling is a repeat of the performance story: the Cortina still feels good, but the Hillman does everything slightly better. Generally, it is lighter, more responsive and less harsh to drive.

IN CONCLUSION Both the new Hillman—in the o/d form supplied—and the well established Ford achieve what they set out to do.

As things stand, the Hillman GT matches or improves on the Cortina GT at just about every point, lacking only the snob appeal that has accrued to the Cortina through Ford's massive and successful competition exploits. This comes as a bit of a surprise but is in part explained by Ford's unfortunate decision, two years ago, to fit to all their Cortinas the bowl-in-piston engine. This meant throwing away the excellent cylinder head design on which the GT's reputation had been built, and with it a good deal of the model's legendary responsiveness and smoothness. Meanwhile Messrs Rootes have only the Marathon win to boast about and belatedly to associate with the GT. In any case, the optional overdrive is essential to bring the Hillman up to the mark and at £55 it is not exactly the cheapest of accessories. To improve the Cortina's potential Ford offer a bewildering selection of engine and suspension tuning equipment. Rootes can offer nothing like this list of extras and the story is repeated among outside go-faster people. Most of them have something to offer for the Ford, while the Rootes range has been ignored except by one or two specialists.

So our conclusion seems to be that the Hillman is the better car. But what good is that going to do Rootes? They've just not got the Ford image. ●

USED CAR TEST

No. 364

1969 Ford Cortina 1600E

PRICES

Car for sale at Shepperton, Middlesex at	£895
Typical trade cash value for same age and model in average condition	£675
Total cost of car when new including tax	£1,097
Depreciation over 2½ years	£422
Annual depreciation as proportion of cost new	15½ per cent

DATA

Date first registered	22 December 1969
Number of owners	1
Tax expires	30 November 1972
M.o.T.	Not yet needed
Fuel consumption	22–27 mpg
Oil consumption	Negligible
Mileometer reading	34,049

PERFORMANCE CHECK

(Figures in brackets are those of the original Road Test, published 28 December 1967)

0 to **30** mph	**4.5** sec (4.1)
0 to **40** mph	**6.8** sec (6.1)
0 to **50** mph	**9.6** sec (9.1)
0 to **60** mph	**13.7** sec (13.1)
0 to **70** mph	**18.6** sec (17.8)
0 to **80** mph	**27.3** sec (26.6)
0 to **90** mph	**46.4** sec (38.8)
In top gear:	
20 to **40** mph	**10.3** sec (11.5)
30 to **50** mph	**9.7** sec (10.0)
40 to **60** mph	**9.8** sec (10.4)
50 to **70** mph	**11.1** sec (12.1)
60 to **80** mph	**15.1** sec (13.7)
70 to **90** mph	**28.0** sec (19.4)
Standing ¼ mile	**19.5** sec (18.8)
Standing Km	**37.1** sec (35.5)

TYRES

Size: 165SR 13 Goodyear G800 tubeless
Approx. cost per replacement tyre £10 (tubeless)
Depth of original tread 9mm; remaining tread depths 4mm right front; 6½mm left front; 3mm on rears; 2mm spare.

TOOLS

Jack, wheelbrace strapped in boot. Handbook in car.

CAR FOR SALE AT:

D. A. Skeggs Ltd., High Street, Shepperton, Middlesex. Tel: Walton-on-Thames 40281.

RE-ACQUAINTANCE with some of the Mk 2 Cortina range now that the current car is familiar is, perhaps surprisingly, refreshing. This is particularly so when it comes to driving a 1600E again. Despite the hint of pretentiousness implied by the "E" suffix, the car does not look as if it is trying to be anything it is not. Appearance is neat, well-proportioned and apart perhaps from the black-painted areas on this latter-day example, unaggressive and free from excess vulgarities. This particular car, which has been in the hands of the same owner for two and a half years and just over 34,000 miles, is on the whole in very good condition. Credit for that must go partly to the sellers, who appear to take an unusually large amount of trouble with secondhand cars, and to the previous owner, who seems to have taken care to keep everything in good order.

In effect the 1600E was a Cortina-Lotus without the twin-cam engine, the single downdraught - variable - choke - Weber carburettor'd Cortina GT pushrod 1,599 c.c. engine being used instead. The so-called "GT" character of this unit is the same as ever on this car — willing and eager, not at all flexible so that a more than average amount of changing down is needed in slow-moving traffic. It encourages the enthusiastic family car driver to "have a go", and is disturbingly noisy at its top end making the car less pleasant than it might be on motorways. Going by the top gear performance figures rather than the standing-start ones — which are not really comparable with Road Test ones since for obvious reasons we do not submit used cars to such severe starts from rest — the only signs of the car's age are at the top of its range. Up to 70 mph it is actually appreciably faster than the Road Test figures. Oil consumption is what one always used to expect from this engine — negligible, although the Road Test 1600E used an uncharacteristic pint every 650 miles. Fuel consumption can drop to around 22 mpg without great difficulty, though up to 26 mpg overall is obtainable; this is typical for this engine.

It is on winding open roads that the 1600E is at its best. The steering is certainly on the heavy side when manoeuvring, but in spite of some intitial slop for which one must compensate it commands the car well. There is little roll for most sorts of driving, and one finds oneself hurrying through bends in great style when opportunities are offered. Handling is good, with predictable understeer normally, and a useful reserve of tail-out oversteer in an emergency swerve, which is easily held. Sidewind stability is not good, and straight stability in still air is average. The ride is adequate, though sacrificed somewhat to handling; dampers showed no marked signs of inefficiency.

Inside the car one finds an unusually full set of instruments, a good driving position, firm but adequate seats in front, and rather less than adequate room behind. One of the biggest points in favour of the Mk 2 Cortina is its ventilation and heating which is first class (though not quite as versatile as the original Cortina "Aeroflow" system). This test took place during the recent spell of very hot weather, and it was a relief to drive the 1600E after some much more modern cars. The bull's-eye ventilators admit welcome streams of fresh air even at low speeds, so that coupled with the air entering via the shut-down heater itself, the inside never feels stuffy.

Brakes work perfectly satisfactorily, and the gearchange is as good as ever. Overall, driving this 1600E was more like driving a nicely run-in new car than one which was first put on the road in 1969.

Condition Summary

Bodywork

The sellers explained to us when we collected the car that they had had to "spray-in" a small scratch on the driver's side of the body just ahead of the windscreen, but it is impossible to detect. Only after some time did we notice that the nearside wing has been resprayed some time before; finish is excellent but the handsome metallic dull-gold paint does not quite match the rest. Underneath the nose, and out of sight for most of the time, the front valance shows a little scratching and pimpling, and is slightly bent. There is a very small "ding" on the passenger front door, though it only shows when light is reflected off it. Some slight pimpling under the paint is evident where stones have attacked the wing returns. One small scratch is to be found on one corner of the front bumper.

The inside of the boot is in extraordinarily good condition, bearing in mind that on one side its single skin in unprotected. The spare wheel still has its cover. Quite a lot of polished wood adorns the interior, and all of this is free of scratches of any sort, suggesting that either the owner was unusually careful or that — less likely judging by the condition of the rest of the car — the wood has been renewed recently.

Equipment

The rare used car which, like this one, is in really good order, makes one look all the harder for defects. The speedometer reads 31 mph at a true 30 mph and 94 at a true 90, the rev counter is (unusually) spot-on and the only item not working was the slow-speed setting of the blower fan.

Accessories

Specification of the 1600E is quite comprehensive, so that little in the way of extras is necessary. Someone had fitted an interior thermometer which gave misleading readings. Door edge protectors of an unusual type, chromium-plated, are fitted, and so is a plated exhaust tail-pipe.

About the Cortina 1600E

It first appeared just before the 1967 Motor Show, a four-door Cortina similar to the GT but with lowered suspension similar to that of the Cortina-Lotus and better equipped inside. One recognition feature was the pair of spot lamps permanently switched into the high beam circuit, giving in effect a four-headlamp system. In November of the following year black rear panels were introduced, the model continuing until being superseded by the current Cortina in August 1970. □

Above: Exterior condition of this Cortina is unusually good, and typical of the car as a whole. The spotlamps are, as usual on this model wired into the high beam headlamp circuit

Left: Reflections visible in this photograph on the dashboard wood give an idea of the excellently preserved state of the interior

Below: Underbonnet accessibility is good. The tubular exhaust manifold is one reason for the 1600E's good performance

Ford 1600E

An inexpensive future classic for those in the know? We discover the joys and failings of this well equipped, sporty saloon from Dagenham.

Is the 1600E a classic?

IF your definition of a classic is a car which offers a touch of individuality, and which marked a significant stage in any one motor manufacturer's history, then surely this Ford must fit the bill? In many ways it marked a turning point for Ford, by showing that they could offer a package of style, luxury and performance in the volume market. Many armchair critics said it just wouldn't sell – but they were wrong! Add to this a production life of only three years, and you can readily appreciate why it has become something of a cult car. An inexpensive future classic, in fact.

So what have we got? Basically, of course, a Mark 2 Cortina body shell blessed with looks which have not dated. A 1599cc four-cylinder engine with crossflow head, bowl-in pistons and Weber compound, twin-choke carburettor, producing 88bhp at 5,400rpm on a 9·0:1 compression ratio. Overall weight is 18½cwt which allows a top speed of close on the "ton" and a 0–60 time of 11·8 seconds. One gallon of fuel is consumed approximately every 25 miles.

To this we can add the lowered and stiffened Lotus-Cortina suspension (independent coil front, ½ elliptic rear) and the handsome chrome and black 5½J sculptured wheels. Then we have all the other features, aimed at creature comfort level, which permits the use of the E, for Executive label. There are comfortable reclining front bucket seats and contoured rear seats with a central armrest. A polished wood fascia panel with full instrumentation and glove locker, and those excellent airflow-type ventilators, give a real feeling of "upper-crustness". The wood continues along the door cappings, there is a padded arm-rest for the driver, which contains a storage facility for odds and ends, and a clock at the forward end of this centre console. A simulated leather-covered alloy steering wheel lends a suitably sporting flavour and acoustic insulation and cut pile carpets help create a driving environment that is a notch above those in "mere" Mark 2s. Externally, twin spots switched into the main beam circuit, black painted radiator grille (and later the back end valance), badges and coach stripe, complete a package which would have set you back £982.2s.1d inclusive of delivery and purchase tax in October 1967.

On the road it is very much a car which enjoys being "driven". It is willing and eager, though not very flexible, needing use of its excellent gearbox in traffic. Performance is quite brisk, most of the power coming in over 3000rpm, but around the 70mph mark the beast betrays its executive label by becoming somewhat noisy. Handling is excellent and one of the main appeals of the car; there is normally a predictable amount of understeer developing into controllable roll oversteer. The car is, however, susceptible to sidewinds, though this failing can be found in cars many times more expensive.

What are the trouble-spots to look out for when investigating one of these cars?

Body

If my memory serves me well, the biggest problem associated with this model has never before been mentioned in this series – namely **Theft.** It is a sad fact of life that "E's" can lose their wheels and interior trim in less time than it takes to fillet a haddock. So be warned!

If the "fly-boys" don't get it, the rustbug probably will. There is absolutely no truth in the rumour that the 1600E was better rust-proofed or used better steel than its Mark 2 sisters. A favourite area is around the headlamps, top of the front wings fore and aft, and the rear outside edge of the front wing panel. Any more than superficial rust here will mean, sooner or later, a new wing panel, and that will set you back nearly £30 for the panel alone. Other obvious areas are the door bottoms and sills, and around the rear wheel arches. Check the jacking points too, because it can be damaging to the car if these give way, and, more seriously, to you if you

Above, plenty of space under the bonnet for easy maintenance. Left, there was a wheel arch here, once. Below, special 1600E name tag.

Ford 1600E

happen to be underneath at the time. Another popular place is the inner wing around the top mounting of the front suspension struts. New plates can be welded in, but they must be **welded** or they will not pass an M.O.T. Another potential M.O.T. failure area is the chassis end-piece for the rear spring hangers, and this too is a welding job to put right. Look out as well for the outrigger bracing members under the front doors -- the remedy again is the welding torch if these are bad. Be careful to watch for body panels which have been filled with "bodge" and sprayed over. Bring a magnet along with you when you look at a car, and remember that this is only a short-term remedy (usually done to sell a car) and that rust will continue to work at the metal around the filler until it falls out, leaving a hole where you don't want one. Fortunately body panels are available from your friendly Ford dealer, and a rear quarter panel will set you back some £30 with sills around £8 each by way of example. Not so easily obtainable are the specialised 1600E interior trim bits. So buy an alarm!

Other bits

Another plus point for 1600E ownership is the engine, which is normally very reliable. Spares are cheap too. A tired engine will display timing chain and valve gear clatter, and other problems will be common to any other engine in this condition; ie, check the exhaust for signs of oil burning. The gearbox gets sloppy as the years go by and may buzz a bit. Clonking in the transmission will indicate that the years have taken their toll here too.

More serious is wear in the steering and suspension department. If the front struts get softer and the bottom ball joints become a little worn, the edge will be taken off the handling; if the steering box has developed some slop too, the car may begin to feel like a mobile blancmange. When you are looking over the car, give the four corners a bounce to check for evenness, or otherwise, in the shock absorbers. Check the state of the suspension swivels with a large screwdriver by levering the suspension links up and down. And finally ensure that there is not more than an inch or so of free play at the steering wheel, otherwise you will be forking out for a failure certificate at the next M.O.T. test.

Finally, a group of enthusiasts who know a classic when they see one, have formed themselves into an owners' club, and you can contact them through John Danvers, 17 Hospital Lane, Blaby, Leics. ●

Above, neat and sporty looking interior, the 1600Es hallmark.

Above, corrosion has started on the suspension flitch plates, and rusty drainage channels will soon be pouring more water over this vital area. Below, aft end of front wing is a favourite place for rust.

Rostyle wheels lend a touch of glamour to the commendably restrained styling which still looks good today.

CARS TO KEEP Ford Cortina 1600E

Dagenham's Delight

It only stayed in production for three years yet most people think of the 1600E as 'one of the best cars Ford ever made.' We ask why

Remember Ford of Britain's 'E' era? It started back in January 1965 with the announcement of the Executive Zodiac and lasted until the demise of the Mk III Cortina 2000E a dozen years later. And it all stemmed from Ford's discovery of the 'executive', a new class of customer visualised as 'a successful young married businessman seeking a four-seater spoting saloon with more comfort and status than is normal for this class of car'.

With the Zodiac, the Ford planners decided 'to build in so much luxury that there really wouldn't be much scope left for optional extras', and specified hide-upholstered reclining bucket seats, automatic transmission and push-button radio for the Executive, which was such an immediate success that it soon accounted for 25 per cent of the Zodiac market.

Exactly two years later, Ford tried the 'E' formula on the Corsair, in a similar fashion to the Mk III Zodiac styled by Roy 'I have so much talent I frighten myself' Brown ('banished' to Ford of Britain after having designed the Edsel). The product planners were anxious to see whether the formula that had worked so well on the Zodiac, selling at well over the £1000 mark, would still prove successful on a lower-priced model. And it did, with '2000E' sales representing 40 per cent of 1967 Corsair production (though I remember the 2000E I drove in 1968 less for its vinyl roof, 'unique radiator grille' and wooden facia than for the welcome relief it represented from the 1.7-litre V4 Corsair it replaced, with its suicidal understeer and rough-running engine with a tendency to oil its plugs in traffic).

So in October 1967, Ford went all the way and introduced the 'E' concept to its best-selling Cortina range, which had just been revitalised with the introduction of the new bowl-in piston crossflow Kent engine in 1.3 and 1.6 litre versions. Only the larger of these two power units was used in the new 'luxury high performance' model, (christened naturally enough the 1600E), which was intended to sell in the 'medium luxury class', a marketing conceit solely invented by the Ford planners.

'We created the 1600E to fill a gap,' recalled a Ford manager. 'We wanted a four-door model that would fit between the GT and the Lotus, offering sporty performance without complications. So the twin-cam engine was out, because of the possible warranty costs.'

Powerful pushrod engine

In fact, the power unit fitted to the 1600E was the most powerful version of the new pushrod engine, as fitted to the GT. With a 9:1 compression ratio, the crossflow engine developed 75bhp at 5000rpm in cooking form but on the GT and 1600E, which boasted a twin-choke Weber, this was boosted to 93bhp at 5400rpm; the performance engine was far torquier than the standard offering, too, and the GT and 1600E could accelerate from 0-60mph in 11.8sec and touch 96mph. In contrast, the 1600 Super Cortina took 16sec from 0-60 and had a maximum of 86mph.

But the 1600E was better equipped to handle its performance than the GT, for it had the lowered suspension of the Cortina Lotus, with an anti-roll bar at the front and radius arms at the rear; it rode on wide-rim (5½in) 'sculptured racing wheels' with 165 × 13 India Autoband radial tyres — crossplies were still standard footwear on run-of-the-mill Cortinas. The chromed 1600E wheels were probably the most expensive fitted to any Ford up till then, but they certainly set the car apart from its siblings; inevitably, after the first year there was a 'cost reduction programme' and the chromium was replaced by aluminium paint.

The performance aspect was further emphasised by twin grille-mounted driving lamps and a leather-rimmed, wood-spoked steering wheel, plus the full instrumentation that was such an excellent feature of the GT, with such 'long-gone' (for Ford, anyway) instruments as an ammeter plus oil pressure and water temperature gauges alongside the easy-to-read speedo and tachometer.

Here the luxury side of this Jekyll and Hyde car reasserted itself, for the instruments were set in a walnut veneered facia (with matching door cappings) rather than the bland black plastic of the GT.

The 1600E also boasted a black cut-pile carpet and extra sound insulation (though *Autocar*, which noted a 'loud exhaust-induced boom', remarked peevishly 'the noise level is higher than is general for this class of car'. Reading between the lines, however, they seem to have driven the 1600E in true 'boy-racer style'!

'The engine is very willing to spin up to the 6000 red mark on the rev-counter and beyond,' wrote *Autocar*, adding: 'the close and even spacing of the ratios (35,51 and 73mph

Note the matt-black rear panel of this 1600E. Not every 'E' had one

The well-equipped interior is one of the 1600E's main attractions

maxima) make it easy to keep the engine 'on full song' during hard driving . . . one seems to be invited to drive enterprisingly as soon as the unit is warm . . . '

Certainly the combination of luxury and performance appealed to the customers. When the 1600E was first shown publicly, at the 1968 Earls Court Motor Show, so many orders were reported to have been placed that the planned production schedules were doubled.

The 1600E attracted export orders, too; production of left-hand drive versions began at Dagenham in November 1968, two months after the first rhd model left the line. At the same time, Knocked Down kits were shipped to Ford's Amsterdam plant to be assembled for the Benelux countries and Switzerland.

Rare two-door version

And it was for the export markets only that the rarest version of the 1600E, the two-door, was created. It made its debut at the beginning of 1969, to coincide with a 'realignment' of the whole 1600E range. The Capri — 'the car you always promised yourself' — had appeared in 1968, and its seats were adopted for the 1600E. More generously padded than the rather flat seats fitted to 1968 models, they held the driver and passenger rather better when they were indulging that 'performance-plus', though the fatter seats also meant that the lidded storage box between the seats of the earlier models had to go. Some consolation was the provision of a fold-down armrest for rear seat passengers, who now also had bucket-style seat backs.

It was then, too, that the expensive chromium-plated wheels gave way to aluminium paint, a black-painted rear panel added, an improved design of central console was adopted, with provision for a push-button radio (the 1968 models just had the radio, when fitted, hung below the dash like the afterthought it probably was) and the dash layout was revised. The gearbox gate was also changed, with reverse left and forward — on the first models it had been right and back.

With the new models came an increase in production rate, and the output roughly doubled. Just over 12,000 1600Es had been built in 1968; the following year, 23,338 were produced at Dagenham and Amsterdam, though only 1872 of these were two-doors (or 'Tudors' to use Ford's nomenclature, created in the Model T age and still current today).

The 1600E certainly fulfilled all the marketing planners' expectations. When production finally ended in August 1969 to make way for the fussy Mk III range, 58,582 1600Es had been built. That represents between five and six per cent of total Mk II Cortina production, making the standard 1600E 'Fordor' an exclusive, if not particularly rare model. Certainly it was the secondhand model everybody wanted when the Mk III came in, for it had a degree of style and opulence the later model was lacking.

But then the Mk III was Ford of Britain's last defiant stand against the obvious merits of 'Europeanisation' of its product range; some senior management were so enamoured of its 'Coke-bottle' profile that they tried to hang on to it into the Mk IV Cortina era. A hideous prototype was built, with the Mk IV's clean front end grafted on to a Mk III backside, but the reaction to it was so hostile that the coke addicts were forced to admit they had been wrong.

In fact, I suspect that the car-buying public would have been quite happy had the Mk II been succeeded directly by the Mk IV, for the later car seems a logical development of the Mk II, both having clean and uncluttered body lines. That could hardly be said of the Mk III, whose sharply-creased bonnet sides were reputed to be so difficult to form that generous applications of body solder were required to maintain their contours.

So the 1600E became a 'classic' almost as soon as it went out of production. The design has held up well, and the four-figure membership of the 1600E Owners Club attests to its continuing popularity. Obviously any 1600E rates as a 'collectable' car these days, but if it's rarity you're after, a trip to the Low Countries could be worthwhile. Surely a few of those 2749 left-hook two-doors survive over there?

I tried to find out whether this successful 'speciality car' had been the brainchild of any one Ford manager, but it seems that it was very much a 'committee car'. Often, that's a sure-fire recipe for a dromedary where a racehorse was intended but in this case the planners created that rare beast, a committee car that happened to work . . . and work well.

David Burgess-Wise

Look closely — this is a two-door 1600E. 2749 examples were made for export between 1969 and 1970

The 1600E's celebrated engine should pose few problems. Accessibility is first class

Brief specification

Engine	In-line 'four'
Capacity	1599cc
Bore/stroke	81mm x 78mm
Valves	Pushrod ohv
Compression	9:1
Power	93bhp (SAE) at 5400rpm
Torque	102lb.ft (SAE) at 3600rpm
Transmission	Four-speed manual
Top gear	17.1mph per 1000rpm
Brakes	Front discs/drums rear
Front sus.	McPherson struts, coils, anti-roll bar
Rear sus.	Live axle, leafs, radius arms, telescopic dampers
Steering	Recirculating Ball
Tyres	165-13 radials
Length	14ft 2in
Width	5ft 5in
Height	4ft 8in
Weight	18.4cwt

Performance

Max speed	98mph
0-60mph	13sec
30-50 in top	10sec
50-70 in top	12sec
Fuel con.	25/27mpg

Production history

Production of the Ford 1600E began in 1967 and the car was launched officially in September that year. August 1970 marked the end of the 1600E in manufacturing terms, by which time a total of 58,582 examples had been built.

No body or mechanical changes occurred during the 1600E's 36-month production run and the only 'update' as such took place in November 1968 when a series of cosmetic improvements made the 1600E's interior slightly more habitable. Our record book states that the rear seat was altered to a twin bucket design, incorporating a centre armrest, a revised facia appeared with ammeter, oil pressure, fuel and temperature gauges housed *in* the facia instead of on top of the dash, and a vanity mirror and rear package tray were also added to the specification. On the outside, the rear number plate panel was now finished in matt black.

Production figures released from Ford show that 2882 four-door 1600Es were built in 1967. 12,057 were made in 1968, 21,516 in 1969 and 19,378 in 1970, making a total of 55,833 cars. Two-door 1600Es became available (for export only) in 1968, and 186 of them were produced in that year. 1872 two-doors left the factory the following year but 1970's total only reached 691 cars. If you're good at maths, you will now have deduced that a total of 2749 two-door 1600Es were produced between 1968 and 1970 which therefore brings the 1600E production total to the aforementioned 58,582 cars.

Buyers spot check

If you're thinking of buying a 1600E, you'll be pleased to hear that the model *per se* does not suffer unduly from the effects of the dreaded rust bug. There are, however, one or two places that should be checked over carefully and the front edge of the bonnet is a good place to start your inspection as the metal behind the chrome strip on its leading edge has been known to rust in the past.

While you're around the front of the car, check the front valence for 'peppering' and the body surrounding the front lights for rust. Next, take a close look at the inner front wings while paying particular attention to the sections which support the bonnet hinges and the MacPherson strut top mountings. Check also the upper edge of the inner wings (where the outer wings overlap) and the area where they join the bulkhead. It's a good idea, at this point, to pull down the sound deadening material from the front passenger compartment, up inside the car; if the inner front wing/bulkhead section has 'gone', a cascade of rust particles will come a-tumbling down!

Moving back along the car, examine the windscreen pillars and the sills (especially at the back) for signs of corrosion. 1600E doors do rust — although replacement skins are available — along with the rear wheel arches but the floorpan should not be a problem. Watch out for rust also around the spring hangers, chassis rails and rear wheel arches (inside the boot), but corrosion of the inner sills is important, though, as these are load-bearing and therefore an MOT fail point.

The five bearing 1599cc GT engine fitted to the 1600E should be OK for a good innings and 100,000 miles could be possible in the right hands. Piston slap and tappet rattle are the usual symptoms of a worn engine but these are not particularly serious faults. Problems with the gearbox and rear axle, likewise, are *very* rare but checking the steering box and its attendant links for wear is a good idea. If the steering feels sloppy or generally show signs of old age, alter the steering box adjustment nut only as a last resort!

Should the footbrake pedal seem to go down a long way during your road test, don't worry too much as this is a characteristic of the self-adjusting rear brakes. The handbrake lever could also be affected in this way so don't be too surprised if the lever comes up a long way. Cars with servos should be better in this respect. The front suspension should be comparatively trouble-free apart from normal wear and tear although the ball joints at the bottom of the MacPherson struts do wear periodically; at the rear, the suspension radius arms have a tendency to clonk when their bushes become tired.

So far so good: but how can you recognize a genuine 1600E? An original car should, of course, wear the distinctive Rostyle wheels which we have heard are no longer made by Rubery Owen, and have the luxurious interior trim described on pages 16 and 17. A 1600E should have front strut strengthening plates on the inner front wings, five-leaf rear suspension (GTs have the four-leaf type) together with lowering blocks between the leaf springs and the axle. Not all 1600Es have these lowering blocks, it must be said, and not every car will have a matt-black rear number plate panel but according to John Danvers of the 1600E Owners Club, every 1600E should have a 'firm' ride. If you are in any doubt as to a car's authenticity, check up with the club.

Don't forget that some 1600Es were sprayed with Ford's notorious 'Silver Fox' and 'Blue Mink' paint colours which peeled very badly when new. But perhaps the biggest worry to an owner of any half-respectable 1600E concerns theft. It is incredibly easy to break into a 1600E (John Danvers said that if you sneeze next to a 1600E the door opens automatically!) so some form of burglar protection, such as a mortice lock on the driver's door, would seem essential.

Rivals then and now

The 1600 was a remarkable car in several ways. It provided performance, luxury and exciting road

Watch out for rust in these areas! Theft is a serious 1600E problem so fit a second lock to the driver's door

manners and all at Ford's customary bargain price. In its lifetime, it competed successfully with the BMW 1600s, Fiat 125s and 2-litre Vitesses of this world and in value for money terms, left them all standing. All three cars had similar performance capabilities to the 1600E but the Ford undercut them all, except the Triumph, on price.

Thanks to its well-equipped interior and stylish overall appearance, the 1600E was verging on Daimler V8, Jaguar 240 and Lancia territory while rivals nearer to home included the V4-engined Corsair 2000E and Cortina-Lotus. Peugeot's 404KF2 saloon was a similar, though more up-market, car in the 1600E idiom but it was much rarer and more expensive than the Ford.

People have been bemoaning the loss of the 1600E ever since it went out of production because it was a fine car that has never been replaced. Fast, compact, luxurious and reasonably inexpensive cars are a rare breed these days and Dolomite Sprint apart, few modern day cars follow the 1600E formula. The Ford Escort XR3, Fiat Mirafiori Sport and VW Golf GTi are certainly fast, compact and competitively-priced, but they miss out on the 1600E's 'olde worlde' interior which, after all, was one of the car's main attractions in the first place.

John Danvers, PR Officer of the 1600E Owners Club and his immaculate eleven year-old car. See 'Owner View' for the story of this superbly-restored 'classic', which John admits has been a 'labour of love'

Clubs, specialists and books

We would strongly recommend anyone who is interested in buying a 1600E, to join the Ford Cortina 1600E Owners Club, to use its full title.

The club is officially recognised by Ford and has 1300 members spread between 20 local branches in the UK. A club magazine called *The Executive* is sent to members on a quarterly basis and this contains news of club activities, technical aid, general articles, members' advertisements and other useful 1600E information. A large national rally is held once a year (Donington was the venue for two of these affairs) and the 1981 rally which took place at Lilford Park in Northamptonshire attracted something in the region of 280 cars. Due to its relaxed, family atmosphere, Lilford park will stage the 1982 'national' and we're told that this year's successful programme (disco, social evening on the Saturday, fun'n games, autojumble and concours on the Sunday) will probably be repeated.

1600E OC subscriptions stand currently at £3.50 per head and *all* club enquiries should be directed towards John Danvers at 54 Auburn Road, Blaby, Leicester LE8 3DA.

As we have stated, the 1600E is an inherently simple car to maintain so no 1600E 'specialists' as such have started up as yet. Any reputable garage should be able to handle body and/or mechanical work and many of the car's components are shared with other Mk II Cortinas. Some components, particularly trim parts *are* becoming difficult to find, however (centre consoles are now impossible to find apparently) so the 1600E OC have taken steps to initiate a national spares scheme but this project is, at the time of writing at least, in its infancy. Meanwhile, the club have arranged a discount scheme with Lucas Service Parts and if you're really desperate for that elusive part, try the Club Mart section of the *The Executive*.

General reading material on the 1600E is pretty thin on the ground and although a certain Paul Pearson was at one stage said to be writing a Cortina book, nothing has materialised so far. 1600E enthusiasts in the meantime will have to be content with a one-off magazine entitled *Cortina Mechanics*, produced by *Car Mechanics* magazine and workshop manuals from Haynes, Autobooks and from Ford themselves.

Prices

At the time of writing, prices for 1600Es fluctuate a great deal but for around £700/£900 you should be able to find a tidy, respectable example with an MOT. A car in this price range may well be 'semi-restored' in that it could have had new wings and sills fitted but overall, it should be in sound shape.

First class 1600Es in A1 condition tend to be priced around the £1500 mark and *exceptional*, concours cars could fetch as much as £2000 but, it should be stressed, a car would have to be in absolutely outstanding condition to realise that kind of money. 1600Es, it should be noted, are roughly double the price of 1600GTs but even rough Es can make £200/£300.

1600E OC members line up at the 1981 Enfield Pageant of Motoring. The club now has 1300 members

Owner view

"At one stage, my whole life revolved around the car. I was almost becoming neurotic! I used to worry where I parked in case the car was damaged, nobody was allowed to smoke or touch the door cappings while in the car, if it was snowing, I would clean under the wheelarches as soon as I was home and if the roads were icy I would worry in case the car skidded into a wall . . ."

That was John Danvers speaking, and he's the PR Officer for the 1600E Owners Club. But if from reading the first paragraph you reckon he's lost his head, you'd be wrong. It's just that a few years back, he was totally obsessed with his car yet since VBC 372H came off the road in 1979, he admits that he has been 'cured' to some extent.

To be fair, most enthusiasts would probably feel the same way if their 'pride and joys' were in the same condition as VBC 372H, for John's car is truly immaculate and a real credit to his loving care and devotion. What's more it's been restored to *original* specification so the net result is a car that, to all intents and purposes, looks as if it has just rolled off the production line at Dagenham.

The car's history is quite interesting for it's a 1970 model that has done 69,000-odd miles. John saw it advertised in the *Leicester Mercury* back in 1974 for £750 but he ended up buying it for £550. He was VBC's third owner though a photocopy of the original logbok from Swansea showed no record of the previous two owners!

Nevertheless, a bottom half respray took place in 1974 and new front wings and a bonnet followed two years later. New sills, wheel arch plates and an acrilic respray, together with a Ford reconditioned engine, new diff, struts and springs plus Spax dampers were also fitted, for John was really into it by this time. "Much of this work didn't really need doing," he said, "but I *wanted* to do it all and get the car in showroom condition."

John has spent considerable time and trouble protecting the car from the elements. The undersides of the wings and wheel arches have been greased and oiled, the axle has been 'hammerited', the box sections de-rusted and the shock absorbers and radius arms sand-blasted, for example' but it's not all over yet for John plans to fit a stainless steel exhaust and treat the front suspension in the near future.

"The car was and is a labour of love," he concluded, "I've spent far too much money on it, and I just couldn't bear to sell it." Having seen John's 1600E, we know just how he feels. □

CONTINUED FROM PAGE 52

pension could be the very reason why the car is so suited to rough work and has come through victorious in yet another East African Safari. The Cortina GT must have the best balanced handling of the four. In most situations its attitude remains quite neutral and only the most gross hamfistedness could ever get a driver into trouble. The slightly wider track of the Mk 2 Cortinas has improved them over the Mk 1s and *their* handling is legendary. Ride is again sacrificed slightly for handling which is not as good as the Pug or 125. It transmits more thump to the cabin and with the smallest wheelbase of the four the ride is choppy over corrugations or constantly changing road surfaces. Despite all our pleas the spring and damper rates of the Mazda's suspension are still out of phase. If Fiat have worked out the secret of soft leaves and strong shockers for the rear we fail to see whv Mazda can't. There is plenty of wheel amplitude with the Mazda but it would seem uncontrolled, so that over dips or humps the bouncing the passengers receive is most uncomfortable. For normal work the ride is good and soaks up cement road joins with hardly a thump reaching the cabin. For this reason the car is more suited to inter-urban tripping (where most will drive anyway) than open road work. The handling is all understeer, not nasty front under plough but definite understeer which in fact makes the car extremely safe in the wet.

COMFORT AND CONTROL

On all round comfort it is hard to pick one as outstanding. All but the Cortina are excellent. To be honest, for six foot staff members none had really good driving positions. The Fiat and Cortina are very similar in having quite upright seating that sets a tall person high in the car. Both are short on front seat legroom but the Fiat has superbly comfortable seats. The Peugeot would have the best position for a tall person with a deep foot well and seats that adjust well back, the furthest of the four. But if the Pug picks up on seating position it loses on pedal position as the brake and clutch pedals are too far from the accelerator. Here the Fiat is excellent where enthusiastic heel and toeing is adequately catered for. The seats of the Mazda have less cushion height seating the driver lower. But the seat can be set well back for good control. All but the Cortina have fully reclining seats as standard with the Fiat having two methods of squab adjustment for "fine tuning". The Cortina's front seats are forward hinged and when set back the squab comes up and the cushion down which is fine when you get used to it but still doesn't provide good seating.

EQUIPMENT AND FINISH

Here we must make a big noise for the Mazda, for whatever else it may be lacking it's not equipment or finish. The Fiat rates right up with the Mazda and has more in the line of functional innovations where in enthusiasts' eyes the Mazda has incorporated equipment for the sake of it. The Mazda has a radio as standard, automatic aerial and tachometer where the Fiat relies on dots at appropriate places on the speedo. A special column stalk intermittent wiper switch (a la Porsche) combined with a foot switch which automatically brings in washer and wipers for a set number of strokes is an excellent drivers' innovation. The Cortina GT and Peugeot miss the boat on a few accounts such as only having manual washers and single speed wipers where $2500 cars in 1968 should have two speed wipers and power washers. The Fiat and Mazda lead on paintwork and trim finish followed closely by the Peugeot. For a car of the price the Ford's finish is atrocious. It does however have full instrumentation and a tachometer which will appeal to the competition minded.

SUMMARY

The choice is hard. Looking back over our sectional analysis we find the Fiat appearing at the top of the list more than any other. Overall it must rate as the most desirable. But satisfying the desire to own a driver's car will cost $300 more than the GT, 404 or 1500SS. In terms of reliability, longevity and performance the 404 at $2550 is unsurpassed value-for-money. The Cortina GT is excellent performance for money and the Mazda 1500SS will appeal to those swayed by equipment, comfort and finish. Certainly there have been many four car comparisons since we tried four cars so even in specification yet so diversified in character. #

PRACTICAL CLASSICS **BUYING** FEATURE

Buying A Ford Cortina 1600E

Superior Cortinas with performance to match their luxury interiors examined by John Williams.

It would be easy to write off the Cortina 1600E as just another variation on the Mk II Cortina theme with cosmetic features such as chromed wheels, extra lights and a more elaborate and luxurious interior designed to make it appeal to those who would be swayed by ostentation rather than by genuine performance. In several ways the 1600E was a Lotus Cortina without the twin cam engine. Ford aimed the car at companies looking for a car which would set their young executives apart from the sales reps., yet would still be simple to maintain.

On the other hand with a top speed of 96 mph, 0-60 mph in 11.8 seconds and a touring fuel consumption quoted at around 25mpg the 1600E had nothing to hide as far as performance was concerned.

A close look at the interior reveals that it is in fact very well furnished with seats which are good in the early models and very good in the later ones and the range of instruments and equipment leaves little scope for additional accessories. It has been argued, probably correctly, that the Cortina 1600E was the best Ford car of its period and there are those who will say with some justification that there has not yet been a true successor. The 1600E is in a class of its own in terms of luxury and economy, it is relatively easy to maintain, and its performance enables it to hold its own in modern traffic condition making it a genuinely practical classic.

Fewer than 60,000 Cortina 1600Es were made and the best examples command high prices.

The Cortina 1600E was launched in September 1967 and 58,582 cars were built by the time production ceased in August 1970. Exterior features which distinguish the model include chromed (and later painted) Rostyle wheels, twin spotlamps mounted in the grille and a single coachline extending the length of the car on each side. The suspension is lowered and modified as on the Lotus Cortina. The interior is puculiar to this model too with a sports style leather covered steering wheel, polished wood dashboard and door cappings and a full range of instruments.

Production changes

The interior of the 1600E was redesigned in November 1968 when the ammeter, oil pressure gauge, and fuel and temperature gauges were moved from their former position in a pod above the dashboard and incorporated into the dashboard itself. The centre console was also redesigned and the locker between the front seats discontinued. The handbrake (formerly of the umbrella type to the left of the steering column) took its place between the front seats and improved front and rear seats were fitted.

Rarities

Amongst the total of 58,852 1600Es there were 2,749 two door models which were produced for export only from 1968 onwards. Few of these are thought to survive and they are particularly sought after by 1600E enthusiasts.

An even more desirable variation on the Mk II Cortina theme is the Savage which was produced by Jeff Uren's company, Race Proved Ltd. Savage 1 was a two door model developed from the Cortina GT and Savage II from the 1600E, retaining the luxurious interior but using the 3 litre V6 Zodiac engine and greatly modified suspension and brakes. Contemporary reports indicate that this proved to be not only a very fast combination capable of 0-60 mph in 9.2 seconds (maximum speed 107 mph) but also remarkably refined and well balanced. If offered a Savage make sure it is a genuine one and not a home-made special and also investigate the insurance premiums!

What to look for

There are a number of areas in the all steel unitary bodywork where rust is particularly likely to develop and some of these are shown in the pictures. Other areas worth checking are spring hangers, rebound stops, petrol tank, the shock absorber housing turrets under the

Buying A

This is the interior of the early 1600E, note the instrument pod on top of the dashboard and the rather flat front seats separated by a locker.

From November 1968 there were several changes to the interior as shown here and described in the text. The seating was improved . . . .

. . . . including the rear seats which became more individual to offer more support.

rear wheel arches and the floor under the rear seat. The boot channel will often be found to be rusted under the rubber seal. All chassis outriggers should be checked as should the top and bottom edges of the door skins and the panel below the windscreen. Wet carpets may not be due to a leaking windscreen but to a rusted bulkhead allowing water to find its way into the car. The floor above the front jacking points and in the vicinity of the seat mountings is also worth checking.

The five bearing engine is said to be sturdy and capable of at least 60,000 miles between major overhauls and when properly looked after 100,000 miles is possible. Wear in the timing chain and valve gear is likely to be the cause of a noisy engine rather than bearing problems.

Two points relating to the steering deserve mentioning here. Firstly any roughness or clicking from the steering is usually due to the balls in the upper and lower worm bearings breaking up. These are sometimes ignored for too long resulting in the steering seizing up. The other point concerns the nylon bushes in the track control rods which should be checked, and preferably changed, every 12,000-14,000 miles.

It is a good idea to ensure that your proposed purchase is a genuine 1600E right down to the front strut strengthening plates on

Specifications and performance

Engine type —	4 cylinder ohv.
Bore	80.97mm
Stroke	77.62mm
Capacity	1598.8cc
Comp. ratio	9.2:1
BHP (Gross)	93
at RPM	5400
Carburettor	Twin choke Weber
0-60 mph	11.8 seconds
Fuel consumption	25 mpg
Max speed	96 mph

Overriders were optional extras but twin spotlights were standard equipment.

There is a useful amount of boot capacity and note that the 1600E spare wheel has a cover.

ord Cortina 1600E

Panels are said to be the most difficult spares to obtain for the 1600E, but DAK Autos (telephone Luton 416789) who have a particular interest in the Mk II Cortinas are planning to supply remanufactured panels soon.

The cross-flow engine should be good for up to 100,000 miles between major overhauls if serviced correctly and the red zone on the tachometer starts at 6000 rpm.

The brake servo (as shown in this picture) was fitted to export models only.

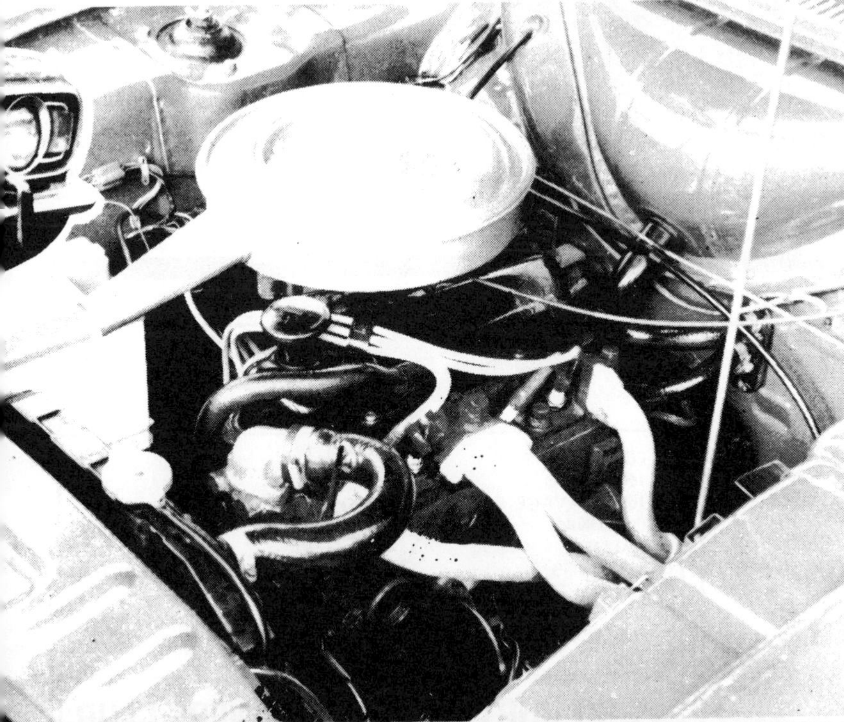

The search for rust damage or evidence that there is rust beneath the paint should start with the area of the front wing above and below the front lamps . . . .

. . . . and continues under the bonnet as shown here. The leading edge of the bonnet is also an area worth checking.

the inner front wings, the five leaf rear springs, and the additional axle location borrowed from the Lotus Cortina; also ensure that the seller is the legitimate owner. These cars are a popular target for thieves and once you own your 1600E you would be well advised to prevent it from contributing to the crime statistics.

Availability of spares

Although the 1600E used the same bodyshell as all the other Mk II Cortinas it is body panels which are the most difficult parts to obtain. A few panels are said to be available

Buying A Ford Cortina 1600E

The suspension mounting areas on the inner front wings should be checked carefully, and you must satisfy yourself that rust damage here has not been disguised and that any previous repairs in these areas have been carried out correctly.

Look at the bonnet channels too and at the tips of the inner front wings which are prone to rust and may go unnoticed being hidden under the channels.

through Ford main dealers although this seems to depend on the attitude encountered at the spares counter — if these panels are available not all dealers will make the effort to find out and obtain them for customers. Scrap cars may still provide a source of trim

What to pay

It is possible to buy a Cortina 1600E for as little as £250 or as much as £2,500, but a reasonably sound example which is neither a concours winner nor suffering from any mechanical or structural problem will probably cost around £700-£900. Give priority to the condition of the bodywork (which will be expensive to repair properly) and the interior which will be difficult and/or expensive to replace or repair.

The vertical rear edge of the front wings is a well known area for rust and the first signs will be one or two small bubbles in the paint.

Examine the inside of the boot carefully paying special attention to the wheel arches, the spare wheel well and the corresponding well on the opposite side.

The sills suffer from rust too, particularly just in front of the rear wheels and the rear wheel arches are especially prone to rust damage.

The Clubs

There are two clubs catering for the Cortina 1600E and further information can be obtained by sending a stamped and addressed envelope to the following:

Mr P, Underwood, **Ford Cortina 1600E Enthusiasts Club,** 54, Fairfield Drive, Dorking, Surrey.

Mr A. Clark, **Ford Cortina 1600E Owners Club,** 22, Stonehurst Road, Braunstone, Leicester.

items but we are inclined to think that you are very unlikely to find a 1600E in a breakers yard with an interior which has not been abused, robbed or completely removed for another conversion of standard Mark II Cortina to 1600E lookalike. By contrast mechnical parts are quite easy to come by. ☐

The writer wishes to thank Roger Viney, Don Raith and John Blackwell of the Ford Cortina 1600E Enthusiasts Club for their assistance in the preparation of this article.

FORD OF EUROPE

CORTINA 1600E & GT 1967-1970

The Cortina 1600E and its sister version, the GT, made a great impression at the end of the sixties. Some 26 articles drawn from the US, Australia and Britain trace their progress over those years. Included are 11 road tests, 2 comparison tests, a 12,000 mile and a 24,000 mile report, 2 group tests, plus stories on their history, continental touring and a useful 1983 article from Practical Classics on buying a used model.

100 Large Pages.

CAPRI MUSCLE CARS 1969-1983

Some 30 articles drawn from Britain, Australia and the US retell the story of the most powerful Capris. Included are 10 road tests, a performance test, a 12,000 m. report, new model introductions, specifications plus a comparison test between an RS 3100 a Lotus +2 130/5 and an MGB V8. Models reported on are the RS 2600, RS 3100, 3000 GT XLR and S, the RSR Turbo, the Black Gold V6, the 2.8i, Tickfords 2.8T, the Comanche 190, Bullit, Chastain S/3, JoMoRo, plus cars fitted with Ford's X Pack also those developed by WM, Allard and Lumo. The 4WD Rallycross 250 bph V6 is also covered.

100 Large Pages.

FORD RS ESCORTS 1963-1980

Rallye Sport Models covered in this book are the Twin Cam, RS1600, RS1800, RS2000 and the Mexico. Some 30 articles trace the development of the RS models from 1968. They include 14 Road Tests, two Comparison Tests vs. the Viva GT and the Fiat 131 Sport, driving impressions, rally reports, touring, new model introductions and full specifications.

100 Large Pages.

LOTUS CORTINA 1963-1970

Stories from Britain, Australia and the U.S. make up the 33 articles that trace the development of the Lotus Cortina between 1963 and 1970. There are 11 road tests, a comparison test, a road report, plus articles on development, racing, rallying, touring, model introductions and testing. Vehicles covered include the Mk.I and Mk.II plus cars modified by IWR, Westune, Taurus, Willment and the works.

100 Large Pages.

These soft-bound volumes in the 'Brooklands Books' series consist of reprints of original road test reports and other articles that appeared in leading motoring journals during the periods concerned. Fully illustrated with photographs and cut-away drawings, the articles contain road impressions, performance figures, specifications, etc. None of the articles appears in more than one book. Sources include Autocar, Autosport, Car, Car & Driver, Cars & Car Conversions, Motor, Motor Racing, Modern Motor, Road Test, Road & Track and Wheels. Fascinating to read, the books are also invaluable as sources of historical reference and as practical aids to enthusiasts who wish to restore their cars to original condition.

From specialist booksellers or, in case of difficulty, direct from the distributors: BROOKLANDS BOOK DISTRIBUTION, 'HOLMERISE', SEVEN HILLS ROAD, COBHAM, SURREY KT11 1ES, ENGLAND. Telephone: Cobham (09326) 5051 MOTORBOOKS INTERNATIONAL, OSCEOLA, WISCONSIN 54020, USA. Telephone: 715 294 3345 & 800 826 6600

FORD MUSTANG

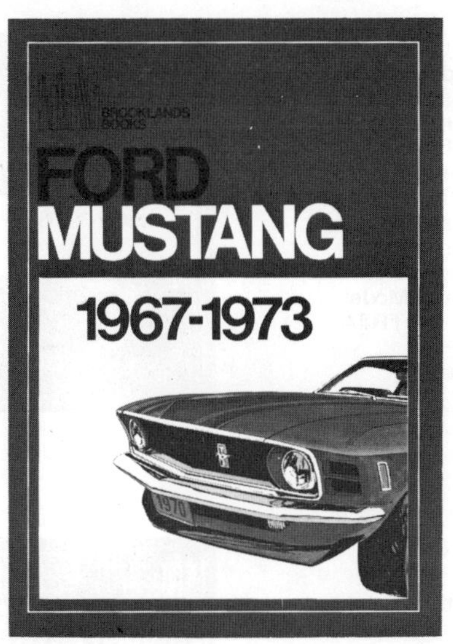

FORD MUSTANG 1964-1967

Road tests, new model introductions, road research reports, comparisons with the Barracuda and Corvair, supercharging, touring, a drivers report and an article on a Wankel powered Mustang make up this book. Models covered... include the 6 and 8 cyl. convertibles and sedans the 350 GT & 390 GT, Ruddspeed and Bertone versions, the Shelby American GT 350 and GT 500 together with 4 articles on the pre-production phototype of 1962.
100 Large Pages.

FORD MUSTANG 1967-1973

A total of 22 articles continue the Mustang story through to 1973. They cover 8 road tests, a track test, history, specifications, an owner survey and comparisons with the Camaro, Javelin, Barracuda, SS454 Chevelle, Duster 340 and the Shelby AC Cobra. Models covered include the Mach 1, Boss 302, the convertible GT, the Shelby and the Trans-Am and deal with the following engines the 289, 302, 351, 390, 427 and 428.
100 Large Pages.

These soft-bound volumes in the 'Brooklands Books' series consist of reprints of original road test reports and other articles that appeared in leading motoring journals during the periods concerned. Fully illustrated with photographs and cut-away drawings, the articles contain road impressions, performance figures, specifications, etc. None of the articles appears in more than one book. Sources include Autocar, Autosport, Car, Car & Driver, Cars & Car Conversions, Motor, Motor Racing, Modern Motor, Road Test, Road & Track and Wheels. Fascinating to read, the books are also invaluable as sources of historical reference and as practical aids to enthusiasts who wish to restore their cars to original condition.

From specialist booksellers or, in case of difficulty, direct from the distributors:
BROOKLANDS BOOK DISTRIBUTION, 'HOLMERISE', SEVEN HILLS ROAD, COBHAM, SURREY KT11 1ES, ENGLAND. Telephone: Cobham (09326) 5051
MOTORBOOKS INTERNATIONAL, OSCEOLA, WISCONSIN 54020, USA.
Telephone: 715 294 3345 & 800 826 6600

TRIUMPH TR6 1969-1976

A total of 11 Road Tests are included in the 26 articles that retell the TR6 Story from its introduction in 1969. Other pieces report on performance over 10,000 and 21,000 miles, comparisons vs. Fiat 124 Spider, the MGB and Porsche 914. Articles also give advice on Buying a Secondhand Model and compare it to the TR4A and TR5.

100 Large Pages.

TRIUMPH TR7 & TR8 1975-1981

A total of 24 articles retrace the TR7 and 8 story from its introduction in 1975. Reports are drawn from Australia, Britain and the U.S. and include 11 Road Tests, a technical and styling analysis, a 24,000 mile report, plus an owner survey, a Comparison Test vs. the Fiat X/19, and articles on rallying and racing the TR8s.

100 Large Pages.

TRIUMPH TR2 & TR3 1952-1960

The TR2 & TR3 story is told through 35 articles and over 200 illustrations. Models covered are the TR2 both hard-top and coupé and the TR3 & 3S. Articles cover road tests, new model introductions, speed trials, touring, development, Le Mans, history, driving impressions and used car tests.

100 Large Pages

TRIUMPH TR4 & TR5 1961-1969

This book deals with the TR4, 4A, 4 IRS, 250, and the TR5. A total of 30 articles from 3 countries covering road tests, competition models, driving reports, new model introduction, race tuning, track tests, Sebring report, owner surveys, race preparation and used car tests.

100 Large Pages

These soft-bound volumes in the 'Brooklands Books' series consist of reprints of original road test reports and other articles that appeared in leading motoring journals during the periods concerned. Fully illustrated with photographs and cut-away drawings, the articles contain road impressions, performance figures, specifications, etc. None of the articles appears in more than one book. Sources include Autocar, Autosport, Car, Car & Driver, Cars & Car Conversions, Motor, Motor Racing, Modern Motor, Road Test, Road & Track and Wheels. Fascinating to read, the books are also invaluable as sources of historical reference and as practical aids to enthusiasts who wish to restore their cars to original condition.

From specialist booksellers or, in case of difficulty, direct from the distributors:
BROOKLANDS BOOK DISTRIBUTION, 'HOLMERISE', SEVEN HILLS ROAD, COBHAM, SURREY KT11 1ES, ENGLAND. Telephone: Cobham (09326) 5051
MOTORBOOKS INTERNATIONAL, OSCEOLA, WISCONSIN 54020, USA.
Telephone: 715 294 3345 & 800 826 6600